Who Dares Wins Publishing
www.whodareswinspublishing.com


First Edition Trade Paperback July 2010 Manufactured in the United States.

eBook ISBN: 978-1-935712-17-6
Print ISBN 978-1-935712-18-3

# WE ARE NOT ALONE
## The Writer's Guide to Social Media

by

**Kristen Lamb**

***Foreword***

I met Kristen in 2008 at the Dallas-Ft. Worth Writers Conference. My Social Media at the time consisted of a sort-of web site and a sort-of blog I occasionally wrote on. She mentioned something called Twitter. Which I'd sort-of heard of.

She helped me get started on a bunch of social media platforms, but I made many mistakes doing most of it on my own and having little clue what I was doing. I wished there had been a step-by-step guide for writers on how to not only do social media technically, but do it content-wise. This book is the answer to that wish.

The first thing I did wrong, was the first thing she teaches in this book. I didn't establish my BRAND. My first twitter name was IWhoDaresWins. I mean. Come on. I had to rebrand myself at Bob_Mayer after a year and a half on Twitter after reading the draft of this book.

Kristen and I have fought over the importance of promotion and social media. I won't say she won, but I have surrendered. (I'm a guy—we never say anyone else won). I believe content is King and Promotion is Queen. And sometimes the Queen rules.

The good news is, with social media, an author can do a lot of promoting from home, although you have to be very careful that you keep it 'social' rather than blatant promoting.

To rule in the publishing world, you need content (king) and promotion (queen). And honestly, my time is split 50-50 between the two. This article is a form of promotion. Every blog I write. Every class I teach. Every tweet and retweet. We are in the entertainment business and a large part of that is selling the product.

Many writers hate self-promotion.

We don't promote because:

We don't want to be considered arrogant.

We don't want to get confronted by people telling us we're self-promoting.

We're not sure what we're promoting is really worth it.

We don't want to be wrong.

Bottom line is it makes us uncomfortable. Under the Myers-Briggs, an INFJ is labeled author. The exact opposite, ESTP, is labeled promoter. Huge problem there.

The first Force of Warrior Writer is WHAT. What do you want to achieve with your promoting? Usually, it's more sales.

I think the second Force WHY, is important because if you understand the real reason why you want to sell more books, it becomes more palatable. For most authors I know, the reason WHY they want to sell books is to make money in order to be able to keep writing. The money is just the means to the end, which is writing. So if we can understand our motivation for promoting, I think we can get more excited about it. Accepting promotion as an integral part of being a successful author is part of the maturation process. The thing is, you don't have years to grow up any more. You've got to do it right away.

Bottom line on this book: While reading the first draft on a flight to present a Warrior Writer Workshop, I realized I really needed to USE this book to revamp my entire approach to Social Media and Promotion.

Every writer needs this book. It will save you time, it will give you focus, and it will help you become a more successful author.

I'm getting the first one hot off the press and I'm going to put aside the time to do everything laid out in it, step-by-step.

***NY Times Best Selling Author***<br>
***Bob Mayer***

Acknowledgements

To my family who believed in me from the very beginning.

I couldn't have ever done this without my mom, who supported me to become a writer from day one. To my brother, Jaysen, who landed me my first job as a technical writer and introduced me to the wonders of MySpace. Bet you didn't see this coming, did you?

To my husband who gave up months of playing X-Box 360 so I could write. Yes, you can go put batteries in the controllers.

Deepest thanks to the DFW Writers Workshop and to Steve Manning and Russell Connor who believed in me and my methods and invited me to experiment...I mean share my teachings with the Workshop. The DFW Writers Workshop is a testament to the power of the collective—individuals helping in small ways every day to accomplish great things. I have been blessed and honored to keep such company.

I am immensely grateful for my closest friends and fiercest allies, the members of Warrior Writer Boot Camp—David Walker, Neena Kahlon, Nigel Blackwell, Terrell Mims, & Dr. Mike Bumagin.

Also, a special thanks to Fred Campos for showing me the wonders of Twitter. Fred Campos, Jeff Posey, and Renée Grosskreutz of Fun City Social Media were always freely available for technical advice used in this book. Thanks for giving so generously of your time and expertise.

Thank you, Bob Mayer, for giving me the opportunity to write this book for *Who Dares Wins Publishing.* I hope it is a blessing to you and many, many other writers. Thank you Jen Talty-Holbrook, *Editor Extraordinaire,* for your hard work and sheer brilliance (Jen came up with the title). I surround myself with people more talented than me. They make me look good.

Finally, thank you, Dad, for inspiring me to love books and to one day become a writer like you. I hope you can see me from up there.

Table of Contents

and rebuilt from scratch. If that isn't a formula to make any writer want to give up, I have no idea what is.

There is a lot of information out there on social media. The problem is that many of the ideas and applications, I feel, work great for Corporate America, but then break down when applied to building a readership. How do I know this? Well, I tested a lot of them. I have joked in my blog that the real title of this book should have been, *"I Made All the Dumb Mistakes So You Don't Have To."* I field-tested a lot of the great ideas proposed in many of the top-selling social media guides. And please make no mistake, it isn't that these books were poorly written or the information was wrong, it is just that writers have different needs than a business or even the casual user.

Not only that, but some of these guides were pretty technical. Before writing this book, I read all of the top-selling social media books available. I've made my living as a technical writer for years, and some of these books were enough to give *me* a panic attack. They were complex, overwhelming, and many of the tools and ideas were of questionable value to a writer trying to build a platform. I witnessed the same play out at conferences. Those holding the conference, in an effort to serve attendees, ran out and wisely recruited firms to come in and teach social media. The problem? Marketing and computer people don't speak the same language as many writers. I could almost watch the audience's eyes glaze over as these high-tech gurus threw everything but the Mashable kitchen sink at them. All I could see was a formula for a writer to be too overwhelmed to even try, or to be spread too thinly and give up.

To me that is tragedy. I want everyone to have a chance to realize their dreams. So, it was time to step up and put together a guide that would be useful to a writer...**all writers.** Published, unpublished, rich, poor. Tech savvy to technophobe. Doesn't matter.

I have created a book so simple that anyone could build a solid platform. You don't need special software or a fancier computer with a super-fast processer to do my program. All you need is an open mind, a servant's heart, a computer and an

## Introduction

Hi, my name is Kristen Lamb, and I am what is called a "social media expert," which sometimes I think is code for, "spends far too much time on the Internet." I wasn't always a social media expert. I started out a regular gal who quit my job in corporate sales to follow a dream of being a writer. When I wasn't writing my novel, I was on the computer because it was the only place I knew to find other people like me. Nerds...no, writers. I meant writers.

I believe that it was in the early days of MySpace that I first saw the potential for what social media could do for an author, especially a new one who didn't have ten grand to drop on a fancy web site. But, I bee-bopped along and didn't too think much about it until a couple of years ago I happened to notice that all my writer friends seemed to be doing one of two things. Either they weren't on the computer at all and firmly believed that the Internet was a tool of the devil. Or, they were on social media, but they were engaging in activities that served little to build their reputation as writers.

*No, I really don't want a digital daiquiri. I already have enough digital drinks to land me in digital rehab.*

These days it seems most writers have discovered the Internet and social media, but I rarely see authors approaching the various platforms with any strategy. They mistake activity for productivity and make rookie mistakes that can cost hundreds and thousands of man-hours to correct. Sometimes the errors are so deep, the platform has to be totally torn down

Internet connection. Everything I teach you in this book has been tested and demonstrated to be effective. Sure the technology is changing faster than ever, but most of the information in this book is written to be applicable no matter what platform you use in the future. There might be gadgets and gizmos I haven't addressed. I have only included the tools and techniques that I have used successfully to build platforms. My tools and techniques are all free, so there is no reason you can't start building your platform right away. Okay, well as soon as you finish this book.

In preparation, I sifted through all the latest and greatest social media books, sites, and gadgetry to give you what I believe works for a writer, and more particularly, the technology-challenged writer. My system, applied properly, will create a rock-solid platform that can support you from unpublished writer to best-selling author.

That and it's super fun.

## Act I—The Big Picture. What Can Social Media Offer?

One thing that happens to so many writers who get on social media is that we can get tunnel vision and fail to look at the big picture. It's an easy thing to do, especially when on the Internet. There are all these widgets and whats-its and shiny things screaming for our attention. Social media sites, in an effort to outdo each other and keep the attention of the increasingly ADD culture, try to offer us everything under the sun. In the end, we often end up using very little, which makes me think of *Hurt Locker.* There is a powerful scene where the soldier, war-weary and home from Iraq, is standing paralyzed in the cereal aisle. The aisle stretches on and on, with endless options, and yet, faced with hundreds of different choices, he cannot make a choice. It's just too much.

There are gadget lovers who know every in and out of Facebook and MySpace, who use it to organize their life, their pictures, hold events, play games, etc. Most of us? I think we figure out a few things we like and stick to them like glue. We chitchat with our friends and post pictures of the kids and every once in a while mention we are writing a book. For the casual user this is just peachy, but for the writer who needs to build a platform and use the most social media has to offer, that can be bad. Real bad. We must get out of our comfort zone. To do this, it helps to get perspective. Not all those gadgets will help you with your big overall plan of becoming a brand and promoting your work. So feel free to ignore most of them.

Now doesn't that feel liberating?

The program I will teach you later is very basic, but guess what? So is a building foundation. Concrete, rebar, sand. Pretty standard stuff. But, when creating a foundation, it is generally a good idea to have a big picture of the building it needs to support. Same with your platform. Let's pan the camera back and get a wide view. Social media is far more than a million ways to blitz about your book. What is the social media experience? Why is it any better than traditional marketing? What it can do for you as a writer?

## Why Social Media?

No, no, no, Kristen! We want a platform. Tell us about Facebook, pleeeaaase!

We will get there. I promise. In my years of building platforms, I have thankfully had my fair share of successes. But, I would be lying if I told you there had never been failures. The blue prints, I think, were solid, but there was one key ingredient that was missing.

Intent. I forgot, albeit briefly, *why* people were on social media.

For you to get the most out of social media, it is fundamental you understand WHY people use it, like it, and are gravitating to it by the millions. If you added the active users of Facebook, MySpace, and Twitter, you would have the third largest population in the world only bested by China and India. There is a reason for that.

So what's the big deal?

Social media is becoming more popular than ever and for good reason. Society is advancing all around us at exponential speed. In fact, everyday life now resembles what was once science fiction. Who would have thought that we would one day rely on phones more than cameras to take a picture? Or that a small box that fits in our hand could rest on a dashboard and give directions?

As our society advances, one of the side effects is that humans are becoming increasingly isolated. We sit in cubicles and stare at a computer screen all day. Many of us even work from home. Most communication takes place via a cell phone and often we don't even get the sound of a human voice. Texting has become the name of the game. Universities are offering more classes on-line, and some are going to the computer altogether. Workplaces are finding that allowing people to telecommute is more cost-efficient. When we bank or pay bills, we often talk to a computer. Buying a ticket at a movie house or even checking out at a grocery store? Yep, also done by computers.

I'm not here to make any kind of value judgment about the continuing automation of our society. That is about as pointless as arguing with a Category 5 hurricane that is making landfall. Rather, my goal is to give an accurate perspective of reality. The reality is that companies are finding more and more ways to replace humans with computers, thus decreasing the traditional forums where we would have historically interacted.

Yet, how healthy is that for the human psyche? When we get down to it, we are designed for camaraderie. It seems that all the creature comforts, while making our lives easier, just seem to make us have to rely on and interact with our fellow humans less and less. We are social creatures who not only desire community we require it to maintain psychological and emotional health.

The good news is that, like water will always find its level, humans will always find ways to be social. We need it. And this is where social media has come to fill the widening gap between humans, by not only providing easy ways to keep in touch with friends and family, but also making it simpler than ever to extend and maintain our network of friends, mentors and acquaintances.

Social media can do a lot for us personally, but what about professionally?

***Let me ask some quick questions:***

What image comes to mind when you think "sales?"

Think back to the last time you interacted with a "salesperson?" Was it a pleasant experience?

Do you feel comfortable "selling?"

Does the term "sales" evoke images of you cousin Marty who hounded you for months about how you could become a millionaire selling vitamins out of your garage?

Have you ever bought a book because someone handed you a free bookmarker? If so, how many books?

Have you ever bought a book off an infomercial?

Does the idea of having to step into the role of "salesperson" sound appealing?

***Don't feel guilty for your answers. But let me ask a few more questions:***

Have you ever told someone about your hairdresser? Good or bad?

Have you ever gone to a movie because a friend recommended it?

Have you ever avoided purchasing a book because someone you knew told you it was a tragic waste of trees?

Would you prefer to skim through a phone book for a mechanic or plumber or ask people you know for a recommendation?

Would you purchase a book in a genre you never read if you considered the author a friend?

Have you ever bought a book simply off name recognition?

It is fairly likely that you answered *yes* to most of this last list. Why? It goes back to that humans being social thing. We are designed to exist in groups, and we have a tendency to trust members of our group before we trust an outside source. If you see someone with a fabulous hairstyle, you are more likely to trust her recommendation than a fancy flyer you get in the mail with a picture of some model and a coupon.

Social media capitalizes on the social *quid pro quo*. Social media recognizes that our world is becoming increasingly technological and that this technology can do one of two things—**alienate** or **congregate**. Social media gives humans the new village marketplace, and those who play by social rules can reap tremendous advantage. Social media makes it easier than ever before in human history to link people with common interests, no matter how far away. Talk about the global village.

The Internet is changing the way people get information. If we want a phone number, we go on-line. If we want to know the movie times, we go to Yahoo movies. Google has become so popular it now has become its own verb. We say "I dunno. Google it."

Until recently, the public relied on gatekeepers to funnel information and discern and guide our tastes. Major networks decided what made headlines and what spin to give the event. Now? News is broadcast often by witnesses on the scene. People "tweeted" information and images about the March 1, 2010 earthquake in Chile before any news crew. If that isn't real news, I don't know what is. Even purchasing decisions are being influenced by social media now more than ever before. Everyone is jumping on the social media bandwagon, from GM to Starbucks. But social media isn't just a trend or a fad. It is a fundamental shift in the way humans interact. Social media is here to stay.

There are a lot of social media books, conferences, workshops, teachers who can teach you all kinds of ways to have a zillion followers and spread your message across the globe.

But let's just take a pause for a second. How many of you have seen the movie *Minority Report*? If you haven't I advise watching it before you get too involved in social media. In the movie, set in the near-distant future, Captain John Anderton (Tom Cruise), while walking through a mall, is assaulted at every turn by flashy holographic ads calling him by name.

*John Anderton. You need a Guiness.*

*John Anderton. You deserve a vacation.*

*John Anderton, John Anderton, John Anderton!*

Now all these flashy holographic ads were calling John Anderton by name, but did that make their approach personal? Or were they just personalized pop-up ads? Your goal with social media is of course to build your platform and create a brand, but we must widen our perspective if we hope to be perceived as better than a SPAM bot.

If we are going to invest our time, then we should do so in a way that makes the most out of our social media experience. We must become involved in our digital community. This involvement will be good for you personally and professionally and will make the critical difference in how others perceive you and your book.

Social media is rewriting the rules of human interaction. What does this mean to you?

## Social Media and Publishing

While most individuals desiring to use social media effectively can benefit immensely from this book, *We Are Not Alone* is designed specifically to help writers build a platform and brand as well as grow and develop their professional networks. And guess what? These tasks might seem monumental, but we don't have to do them alone, ergo the title.

But before we run out and make friends and write blogs, it is a good idea to understand the climate of our professional world.

The world of publishing is not immune to the sweeping changes affecting virtually every type of industry. In fact, some speculate that traditional publishing will disappear in the next decade. I don't know if it will disappear, but it certainly will undergo massive changes. It has to. The traditional approach to printing a book and sending it to market is grossly inefficient and wasteful. Not only that, but we now have entire generations that rely solely on electronic medium. Those are the readers of the future, and publishing will have to accommodate to survive.

The music industry failed to see what was on the horizon with iTunes and Napster. It was a fatal mistake, one the music industry has not recovered from. Record stores have become a quaint anachronism and most major retailers severely limit their inventory of CDs. Unless a musician or band is at the top of the charts, they often don't get shelf space. This drives most music purchases to the Internet and to downloadable format. Music has gone digital.

Photography, too has gone digital. Major film companies like Kodak have suffered tremendously, and it is likely that most photo-processing outlets will fade into the history books in the next decade. Film photography, for the mainstream consumer, is dying. Video is also going solely to the digital format. If you don't believe me, try to buy a camcorder that uses film. So what about publishing?

Steve Jobs, CEO of Apple, announced on May 6th of 2010 that iPad users downloaded 5 million titles in the first 65 days and the growth in eBooks continues to climb. Though the Big 6 of publishing seem to be conducting business as usual, there are all the telltale signs of a major shift in how readers will be getting their entertainment. The iPad has changed a lot of people's minds about how fast paper books will be replaced by digital.

I know some of you are like me and just looove paper books. I write in all of my books and it is harder to write in a

Kindle. But remember we are thinking about strategy here. We don't just plan to captivate the readers of today, we must also plan for the readers of tomorrow if we want a *career* as a writer. Every one of those teenagers or college kids glued to their PDA is a glimpse at our future readers, and we have to anticipate catering to *their* needs, regardless of our own preferences.

I recently attended a conference and one of the agents dismissed the e-book sales as trivial and statistically meaningless. He didn't believe that books would eventually go paperless. He might be right, but in my experience, fortune favors the prepared. Building a great platform via social media can only help your "paper" sales, but it will mean the difference between survival and extinction should books go digital like so many other entertainment mediums.

## There is Good News and Bad News and Great News

### So the world is going digital. First, the good news...

It is easier than ever to download content (music, video, blogs, pictures, etc.). Ten years ago, the Internet was still a tricky bugger. I once tried to e-mail three photographs, and I locked up the server so badly I had to call my Internet provider to take my pictures off the network if I ever hoped to use my e-mail again.

The MP-3 existed (I had the very first one) but there was nowhere to download music, and one needed a Computer Science degree to figure out how to use the darn thing. The vast majority of consumers were terrified to buy anything off the Web for fear evil hackers would steal their credit card number and be halfway to the Caribbean before authorities were even on to them. Not so long ago it was a tough and terrifying world.

But now? Technology is amazingly simple and far more user-friendly. I remember the days when if you looked at a computer wrong, it locked up and crashed. If you went on the wrong site, your computer went haywire, and you didn't know whether to call a technician or an exorcist. Yet, with better processors, firewalls and software, a lot of those problems are, thankfully, things of the past.

Today's technology is so simple a four-year-old can use it. Literally. This increased simplification has opened up new markets never before tapped. Think Grandma and Grandpa. A lot of the intimidation factor is gone from computers, leaving more people willing and eager to embrace the future.

It used to be that only a very slim, very brave, and very wealthy cross-section of people owned computers and actively surfed the web. One had to either be a computer programmer, so that if something went wrong you could resurrect your computer from the blinking-blue-screen-dead. Or, you had to have access to enough money to fix or replace the dead computer. These problems made for a super-narrow computer savvy demographic. Yes, I *typed* all of my college papers on a typewriter because I was definitely *not* part of this cross-section.

Yet, today the barriers are coming down. Technology is safer and easier than ever. "When in doubt, *reboot*," as I always say. This means there is a significant widening of the demographics one can expect to find using technology and the Web (*psst, "widening demographics" is code for "more readers"*). Eight years ago the only reason my mother went near a computer was if she had a question for me...who happened to be working on the computer. My mom was one of those technophobes who believed if she hit the wrong button, she would break the Internet. Now? She pays bills on-line and checks her Facebook on her PDA. Yeah. I didn't believe it either.

The Silver Surfers are taking the Internet by storm. *Silver Surfers* is the term given to a population (usually over the age of 50) spending growing amounts of time on the Internet. This is a very promising development for writers hoping to create a large fan-base. People are living longer than ever, so this population is growing more rapidly than any other. Also, this demographic tends to have more disposable income and time, now that they no longer have young children to care for. This means, more money to buy books and more time to read books. Yippee!

Many companies now are vying for the attention of the gray/silver market, which, for the first time in history is increasing as the youth market is declining. Publishing will obviously want to take advantage of these new marketplaces.

More good news? Technology not only is getting easier to use, but it is also getting *cheaper.* E-readers are expensive now, but give them a few years. It wasn't that long ago I had to practically take out a small business loan for a Blackberry and now they are practically giving them away in Happy Meals. Okay, maybe they aren't *that* cheap, but when cell companies are selling them "Buy One, Get One Free," in forty-two different colors and for a tenth of the cost of five years ago? It sure feels that way.

E-readers are a definite trend for the future. They are portable, lightweight and have the ability to change the font size. That will be very appealing in the days to come. It will not only be less expensive to keep up with a book habit, it will be easier on the eyes and on the back. Books take up a lot of room in a home and make moving tough and costly. Yet, with e-readers, consumers can limit the paper books to favorites and moving all those great romance novels or thrillers is as easy as tucking an iPad into a briefcase or purse. This will make many people more daring and willing to take risks with lesser-known or unknown authors. Good news for you new authors, right?

Another bonus. No longer will font size be a determining factor in a purchase (ahem—Silver Surfers?). I've run across books I wanted to read, but didn't because I didn't wish to go blind in the process. That has all changed now that I have my Nook.

Font size is a purchasing factor and likely why the trade paperback is so appealing. Mass market books can be hard on the eyes, but hard cover or large print books lose portability. With e-books, font size and portability are no longer related.

Another point. Now that the Internet is bringing content directly to the consumer, it will be easier than ever for publishers to determine what is resonating versus what is a flop. This means that new authors will be afforded more opportunity for publication. I predict that publishing houses will sign new authors to e-book contracts with an understanding that success in the e-format will open the gateway to traditional print formats.

**The bad news?**

Publishing houses will likely divert the lion's share of marketing dollars to their top best-sellers at the expense of the mid-list author. New authors will find little to no support. While there likely will be more opportunities than ever before, more and more marketing responsibilities will inevitably shift to the author. Just recently I watched an interview with one of the big publishing gurus who predicted that every book ever published will soon be available in e-format. Talk about competition! As it stands a published author today is competing just with books in print. In ten years, *a writer will be competing against every title ever published.*

Breathe. It will be okay.

Three years ago, when I first started my social media soap box, most agents didn't care if writers had a platform unless they were non-fiction. Additionally, they didn't give a flying hoot about a writer's presence on MySpace or YouTube. They generally wanted to see a platform built via traditional media methods—articles, publications, television appearances, public speaking, etc. If a NF author had a web presence, agents focused on whether that writer had a blog and a formal web site with demonstrated traffic.

That has all changed.

Agents now want to know what sort of social media following a writer can demonstrate—for non-fiction *and* fiction. Publishing is a traditionally slow business, but they are picking up speed.

Are you?

Herein lies the rub. Agents want you to have a social media platform for your content. Just because you have a blog or a Facebook page doesn't mean an agent will take you more seriously. And just because you have a blog doesn't mean you have a platform. This is where the strategies I will teach you will set you apart from the rest.

Authors need more than a social media presence. They need to create a social media platform that one day has the power to generate sales.

Ack! I used the "S-word."

The problem is that authors often have little business knowledge and marketing savvy and the mere mention of the word "sales" is enough to make us break out in hives. Most of us writerly-creative-types are NOT salespeople. I was in sales for years, namely because I was an extrovert, not because I particularly liked selling. Heck, the thought of sales evokes hazy images of childhood trauma, of being forced to go door to door selling half-melted overpriced candy bars for some school fund raiser. Most writers have this shared dream of living in a beach bungalow banging away at a keyboard. We have visions of sending our brilliant manuscripts to New York where the publishing house magically makes us all rich best-selling authors. Sigh.

Yeah...not gonna happen. Sorry. I was kind of bummed too.

But, there is more good news. In fact, it may even be great news. In fact, it might even be the best news you get when it comes to your writing career.

**The great news.**

For the first time in writer history, you actually have some control over your destiny. Remember that second list of questions? What if you could befriend thousands of people? What if all those people associated *your name* with your writing? And what if that network of people could multiply your exposure exponentially? What if, instead of reaching a couple hundred people, you could reach thousands?

Sound too good to be true? Not really. And, no need to panic. I am not dancing around wearing giant sunglasses and a suit plastered in dollar signs.

*You too can make millions selling foreclosed real estate!* Kidding.

I am here to help you use social media effectively, not only for fun, but with purpose to build a platform and ultimately create a brand using relationships. If you are going to spend time building this network, then we need to make sure you are doing it effectively, that you are branding the correct name, focusing on content that serves the reader, and recruiting allies to your team to help you succeed.

Writers often have to get over the angst of having to sell. Heck, for many of us it took years of expensive therapy and often couple of glasses of wine to even have the courage to say, "I'm a writer" out loud. Now we have to *sell* too? Yes, but not really. My goal with the book is to change the writer's behavior, not the writer's personality. I am assuming most of you are not doing this writing thing until your career in sales takes off, right?

No one? Okay, you NF authors always have to be funny. We *know you guys* don't mind selling. But the rest of you? No one?

Good. Because I am going to make this fun. Promise.

## Everyone is a Salesperson

Have you ever eaten at a really good restaurant then later recommended it? Have you ever seen a movie that was so outstanding you badgered your friends and family until they watched it too? Have you ever had a great electrician, roofer, mechanic, or nail technician that you referred? All of us, in some way, have seen or experienced something so pleasant we felt obliged to pass on that information. We, effectively, were acting as salespeople.

I was in sales for over a decade. All kinds of sales. I sold knives, vitamins, water filters, and even newspapers door to door in the worst areas of Dallas Fort Worth. Hey, not like I had a rich uncle to pay for my college. Later, after graduation, I worked in jewelry sales, then technology, and finally landed in international sales as a rep for one of the largest paper companies in the world.

Plain fact of the matter? Most people don't trust salespeople. Heck, even salespeople don't trust salespeople. And before anyone gets in a ruff, let me qualify. The best salespeople make it seem as if we are in charge and that *we* are making the decisions. They *serve* us, they don't *sell* us. I don't care who you are. Most of us do not like to be bullied or manipulated. If manipulated, we might make a purchase, but we are far less likely to feel good about the purchase and even less likely to recommend the experience to others.

Used car salesmen are a stereotype for a reason.

It is this almost allergic aversion to "sales" that has made social media so appealing for the population at large. Social media places the consumer in the driver's seat of choice, and capitalizes on "relationship sales." As a consumer, we can get on Facebook and join discussions about cars, computers, printers, cameras, and day care. No one is making a red cent for telling us that the Whirlpool Washing Machine 2000 is the best thing since sliced bread. The feedback is genuine, good or bad.

Companies are being forced to rely more and more on this type of customer buzz. They have to tap into any negatives to prompt remedy any issues, but they are just as eager to capitalize on the positives. What does this mean? It means companies are being held accountable more now than ever.

What does this mean for authors? It means just what I said. Content is more important than ever. Social media is changing purchasing habits. If GM and Sealy Posturepedic can appreciate this reality, then the wise author should as well.

Ahhh, but here is where social media tactics *must* change for authors. If we befriend Starbucks on social media, we don't really expect two-way communication. So, when Starbucks sends out a blanket e-mail with a coupon for a free Frappuccino, we are thrilled. When an author sends out blanket e-mails and **only** blanket e-mails? We are offended. Writers are different than businesses. We are responsible for content and interaction unless we are so mega-huge that we are the Starbucks of writers. Most of us aren't.

Your "sales" objective with social media is to tap into our collective need to serve others. We all seek validation, belonging, self-actualization. That's why money will only get so much out of us. Humans, for the most part, must have meaning to what we do. We like to help. It makes us feel good. This is why spamming us about your book just makes us feel like John Anderton and the holographic ads. We consumers need to feel as if our buying your book is *us giving to you*, not *you taking from us.* Granted, with luck and hard work there will come a time that you will have to rely on bulk mailings and other impersonal marketing tactics. But, just a small added personal touch will make your blanket marketing far more effective. If it didn't, then Dave Thomas on all the Wendy's ads would not have been the success it was. Consumers like faces.

But, Kristen! How can we ever hope to *manage* a network of thousands, let alone be *personal*? Don't worry. I will show you how to do this over the course of the book. One step at a time.

## Social Media and Connecting with Readers

Tapping into reader and potential reader populations is easier than ever with social media, because of two factors.

First, we don't have a lot of time, and so we often look to what our peers instead of traditional marketing to tell us what is good. Nowadays, instead of consumers relying on what books happen to meet them on the front shelves and tables at a local Barnes and Noble, they are looking to their social networks for the skinny on what books to buy. Why? Because they know that Facebook or MySpace is a far better place to get an honest opinion.

Placards and display tables may not be the most accurate litmus test for a great book. Of course publishing houses will tell readers that Book X is fantastic. They have a sizable financial investment in Book X's success they hope to recover. They are also the ones who will ultimately profit from Book X's sales. So maybe not the most unbiased party. Your buddy Frank, however, makes nothing off telling you that Book X was awesome. He also loses nothing by telling you that Book X is so bad that using it to line a bird cage should be considered animal cruelty.

This is why the "sales" that happen via social media are the most genuine kind. When individuals "tweet" about a great book they just read, odds are they just want to share a great experience with their digital community. This is the kind of "selling" you want for your book and your name, the very kind of promotion that can help you hit gold when it comes to finding future fans. Millions of people are gravitating to social media because of this evenhanded honesty. Bring spam and form-letters into their sacred space, and they are far more likely to resent you than to thank you.

The second reason that social media is invaluable for finding fans is that it is easier than ever to connect to people from all over the world based on common interests. Fifteen years ago, networks were limited by geography. Unless we wanted to travel, write a lot of letters, or pay a hefty phone bill, we pretty much had to hang out with people who were close by. Now? That has all changed. I network with people all over the world every day and for free as if they are my buddies down the street. This is a blessing we need to take advantage of to help build our platform.

People generally like what their friends like. I will teach you how to influence key individuals on your social media network. Rather than wasting time befriending hundreds and thousands of people on social media like some bot, I will give you methods to hone in on key influencers of large groups and develop working relationships.

### The Old Versus the New

I think one huge mistake all of us make is we risk falling back into that old way of thinking about marketing. We believe we must do it all on our own, so we feel propelled to go make hundreds and thousands of friends to be "good, responsible little marketers." That is a major fallacy when it comes to social media. That is **conventional marketing.**

In traditional marketing, a brand was passively received, thus the brand had to be controlled and one-dimensional to keep from confusing the masses receiving the image plastered on billboards, placards, magazines and broadcast on radios and TV. A brand had to be static and fixed because any deviation could confuse the consumer and dilute the message.

*Just Do It.*

Nowadays, branding is highly organic and always in flux, namely because we are in the Information Age. We are constantly being fed real-time images and impressions via YouTube, Twitter, FB, blogs. Not only are we being fed these impressions, but then we often take them in, filter them then recycle/repackage them when we resend them out to our community in the form of our opinions. And this is why our marketing approach must be fluid and dynamic. We want people to take in our message, like it and deliver it to their communities in a positive way.

To accomplish this, our approach mush be modified.

Marketing is now in the hands of the audience. Thus, now it becomes critical what the audience thinks of us, because that will affect how they handle our message.

For instance, 20 years ago, it was far less important whether an author was a nice person or not. Who cared? Could she write? An author could have been the biggest jerk on the planet and it didn't matter so long as she didn't do anything that made national headlines. She could hand in her books, and then the marketing/PR people controlled what impression went out to the masses, if any. Writers could live quiet lives of obscurity, and it really didn't affect their book sales.

Now? What a writer's fans *think of her as a person* influences her marketing. Now, this author can choose to do nothing, and the PR people will keep sending out her crafted image. But what if she wants more? She needs to get in the mix. The more an author interacts with her fan base in a positive way, the more likely those fans will pass on her messages in a positive light. By continual personal and positive interaction, an author can influence groups of people to extend her marketing influence. How? She has recruited her fans and followers to be part of her team. Book sales and promotion have now become a collective endeavor.

With the shift into the Information Age, authors are no longer permitted the luxury of obscurity. Long absences between books might have been standard before the 21st century, but now the modern fan expects more interaction. We consumers are plugged in and want to hear from you. If we don't, we will gravitate to an author who is connecting with us.

Back to my earlier point. We must remember we have others to help us if we will just enlist them. We need to recruit allies to our team.

Traditional marketing was a demographic-numbers game. We threw enough stuff (content, mailers, coupons) at a particular wall (demographic) and then hoped something stuck that generated word of mouth or "buzz" that would translate into readers and then into fans. This tactic works in traditional marketing, but often breaks down on social media. The companies who do the best on social media have a different approach for social media that appreciates the consumer's participation and need for community.

Same with writers.

We do better focusing on authentic, positive relationships and then enlisting others to carry our message. We don't have to have 6000 friends to reach 6000 people. We must have a good amount of authentic interactions with others who will then carry our message to *their* networks. Do this correctly, and eventually *their network* will become *our network* and our influence will spread exponentially. In social media, it works better to employ relationships to spread our message, not fancy software that can target customers and SPAM them.

But, trust me. This method is way more fun anyway.

If you hope to get the greatest benefit from social media, then we need to pan our camera back even farther for the even bigger picture. Other than marketing and chit-chatting with friends, what can social media do for us?

So glad you asked.

**Beyond the Selling Books Stuff—What Can Social Media Do for You?**

Earlier we discussed the tendency to get tunnel-vision when getting on social media. We can get so overwhelmed that we risk catching this myopic perspective of what social media can offer. I see it all the time on comment boards.

**Comment**: *Oh, well no one is really on MySpace anymore, so I keep up with my friends on Facebook.*

**Translation**: *I don't understand the power this MySpace page has and I am too overwhelmed to find out. So I go where I am comfortable and that is with my friends and family.*

**Comment**: *What real value are three thousand friends anyway?*

**Translation**: *I don't see how it is possible to interact with large groups of people and it scares me, so I will stay where I am comfortable.*

**Comment**: *Well, I am not a real writer yet, so I don't see the point in building a platform.*

**Translation**: *I am terrified of failure and if I start building a platform, I am married to this dream for better or worse.*

It's okay to have fears, and all of these comments are common and valid concerns. But, let's push past this idea of platform, brand, platform, brand and get a bird's eye perspective. Social media has a lot to offer all levels of authors. You just have to know what's out there.

## Unpublished Writers

It is never too early to start building your platform and brand. Yet, if you are brand new and maybe aren't even sure what genre you want to write, PR firms really can't help you. But, beyond that, new and unpublished writers are in need of coaches, mentors, and moral support. Writing can be very rewarding, but it can be very lonely, and it is easy to get discouraged or off track with bad information. Also, publishing is not immune to its fair share of predators who will take advantage of naïve new authors. But have no fear. Social media is a great way to learn from the best in the business, from authors to agents to publishers. Sites like Facebook and Twitter make it easy for even the greenest pea to network with people who, in any other place and time, would be virtually inaccessible. Learn the ins and outs and dos and don'ts of publishing from those who have a solid reputation in the industry.

How many new authors are in writing groups comprised of people who have never been published? These groups are valuable, granted, but their advice is limited. That would be like me trying to teach someone to swim competitively using a text book. Wouldn't you rather place your fate in the hands of an Olympic swimming coach?

On social media, an aspiring author can make mentors of their favorite best-selling authors, the best agents in NY, and even make mentors out of the major publishing houses. What a vast resource of information. Instead of wasting time trying to sift through books, blogs and articles on your own, you now can zero in on resources the 5%ers of your industry recommend. Also, the most successful individuals tend to have different life habits than the rest of society. What a tremendous opportunity to "peek inside" the lives of your heroes. What are their daily habits? How many words a day do they write? How do they plan their novels? What research tools do they use? The list goes on and on and on.

As an example. For ages, I had a really difficult time wrapping my mind around crafting a plot. Yet, I follow NY Times Best-Selling Author Susan Wiggs on Twitter, and one day she happened to "tweet" about Jack Bickham's *Scene and Structure.* I immediately ordered the book, and it has become a staple text in my novel writing workshop. Instead of having to sift through countless books on plot and structure, I was able to zero in on the best…because it was recommended by the best.

It is incredibly important at this stage of the game that you, the new writer, do all you can to become good at your craft. Natural talent does count, but writing is tough and there is a lot to learn. For instance, one can have natural rhythm and coordination, but if you want to dance for the New York Ballet, you need to *study* ballet. The better the teacher, the better your odds of nailing the audition for *Swan Lake.* Writing is very similar. Are there people who wrote one book with no instruction and hit it big? Sure. But people win the lottery every week, too, and I would not advise scratch-offs for your retirement portfolio.

The wise writer who wants a career needs to study from those who have been down that same path and who can give her instruction, tips, and advice about the craft of writing. Community college classes, continuing education classes and even returning to college for that MFA are all very limited, time-consuming, and expensive approaches to learning how to write commercially. So if you desire to write commercially, why not learn from those who are already doing it?

While you are busy learning about the art and subsequently crafting your opus, you can also use social media to become familiar with the standards of the publishing industry. This ability to keep a finger on the publishing pulse is becoming even more vital in that, with the influence of technology, the rules and paradigms will be constantly changing, and it will be incumbent upon the writer to keep up. It is your business after all.

Social media is a great way to learn the rules of your profession. I remember years ago (pre-Internet) reading a book that suggested new and different ways of getting attention for a query letter. This book actually recommended colored, scented paper and tricks like pasting the query inside a pizza box. Seriously. I wonder if this book was partly responsible for birthing at least some of the urban legends of misguided queries that went tragically wrong. But how many well-intending authors paid money for a book they *thought* was giving good advice? How could any newbie reading this book know that these weren't acceptable practices?

I remember thinking it was a clever idea, but I was green as grass. I am just glad I didn't have a novel finished at the time, or some agent in NY would have received a pizza box with my query...and I would have had to choose another name or profession.

Give me a break! I didn't know!

Yet, now with the Internet and social media, it is simpler than ever for new writers to learn the correct way to capture the right kind of attention...directly from the source.

Agents and editors often blog about a variety of topics, from writing techniques to crafting the dreaded query letter and synopsis. They post about what types of books they are currently in the market to acquire, vital information that can help you narrow your query search. On Twitter, agents and editors do a lot of free critique and even hold question and answer sessions similar to the format of many writing conferences—only this is FREE. Such insider information makes it easier than ever for a new writer to put her best foot forward when it finally does come time to query.

Additionally, social media helps you keep in touch with other aspiring writers so you can offer each other encouragement and accountability. Following best-selling super duper authors is indeed fantastic, but never underestimate what your fellow authors can contribute, regardless their current level of success. Remember, Stephen King was not always a household name. Yes, even the King was an unpublished writer back in the day.

So feel free to expand your network to all kinds of people, best-sellers and wanna-bes alike. Your social media network provides you a way of making the vast amounts of Internet content manageable. Truth be told, you will quickly see who regularly posts good content, versus those who fill up space with a bunch of fluff.

Ultimately, social media can help get you plugged in. Get the skinny on the best conferences and the right organizations to join to beef up that query and show you take your craft seriously. Your network can also prove invaluable for opening the doors to opportunities like teaching at conferences, writing book reviews, blogging for a community examiner, or even writing smaller pieces for on-line or local publications. Most of my early writing gigs came from businesses and individuals who found me via MySpace.

But most importantly? Social media can gain you a reputation for your content, and link you to others who like your *content.* No time like the present to begin building your fan-base of *readers*. I'll teach more on how to do this later in the book.

### Published Authors

First of all, kudos to you for such an amazing achievement. This is a brutal industry, and if you can count yourself among the ranks of the published, you have already proven you have the mettle to endure. Yet, now that you have a little momentum, you might be finding that it feels like a full-time job to keep afloat, let alone improve from here. The pressure is on. You have demonstrated that you have the capacity to succeed, and now it seems a strange reality that anything other than future successes is often equated with failure.

I know that as new authors, we have a tendency to believe that, "If we can just get published, it will get easier." Yet, it is often just the opposite. Many published authors will quickly confess that the pressure to perform only gets worse with every achievement.

NY Times Best-Selling Author Allison Brennan left this comment on my June 5th blog, *Writers—Do You Have the Right Stuff?*:

> *It doesn't get easier when you are published. In many ways it's harder. Sure, I now get paid to write and I don't have to go to the day job, but it's still a tough business, and even though I love writing, when there is the business of writing involved it makes it hard. Then there's improving on each book (or trying to); reader expectations; editor expectations; lists, reviews, social networking, blogging, and you realize that there's not enough time in the day to do it all and, oh yeah, raise a family, sleep and breathe.*

So, in short, it doesn't get any easier once you are published. You have figured that out about now I am sure. In fact, according to the Book Expo of America, 9 out of 10 first time novelists will never see a second book in print. Getting published is just one step, then sales takes over. Sales numbers have the power to make or break a writer's career, and it breaks a lot of them. In 2009, 93% of published novels (traditional and self-published) sold less than 1,000 copies.

Most first-time novelists fail to sell out their print run. I am not here to depress you. Quite the contrary. Social media, used properly, can help you beat those nasty odds.

Social media can extend the network of people who know you, recognize you, and who equate your name with a product (book) or idea/content (romance, thriller, Nazi history, etc.). Thirty-five years ago, people might have said, "*Stephen King who?*" Now, Stephen King is synonymous with horror. That is every author's big long-term goal. You might be published, but does the average reader put your name as synonymous with your genre? If not, then social media is one tool to help that happen a bit faster than it did for Mr. King. He had to bust his *tuchus* for over a quarter of a century. Hopefully social media will hasten this process.

So, social media can multiply your efforts in becoming a brand. We'll tackle that in a bit. What else?

I believe that most of you will admit that, just because you are published doesn't mean you know everything. Like the new author, you too will benefit from socializing with other authors, with best-sellers and even keeping tabs on what's happening in the publishing industry. Your network will make it easier to hone in on the best blogs, articles and web sites to learn more about your craft or even keep tabs on what's going on in publishing, better ways to sell, promote, market and engage your readership.

Maybe you like your agent, but maybe not. What a great way to get to know an agent beforehand. Are they kind, professional, hard working? Social media can provide priceless "intelligence" about agents, editors, publishers, critics, etc.

What about opening the door to new opportunities? Unless you have peaked I am sure you would like to stretch your wings and take a chance at taking your career to a higher level. Surely you have heard the adage, "It isn't what you know it's who you know." That axiom certainly holds true when it comes to publishing. Humans default to who they know first, then look to strangers as a last resort. Social media can help position you to be in front of the most doors of opportunity.

Picture it this way. You are standing in a room of **three** doors (doors representing opportunities). Now you are standing in a room of **thirty thousand** doors. Which room do you believe will provide you with the greatest odds of a door suddenly opening? Now which room do you feel has the greater likelihood that, not only will a door open, but it will be the pivotal door that changes everything about your career as you know it?

Success is all about playing the odds, and social media drastically shifts the house odds (as in *publishing house odds*) to your advantage. The larger your network, the greater the likelihood of success.

Before I teach you all about search engines and platforms and marketing plans, it is crucial to understand some fundamentals. My goal is that by the end of this book, you will not only know **what** to do, but also understand **why** you are doing it.

## Marketing 101

People have told me time and time again that it is impossible to market fiction. I don't believe in impossible and I bet half my motivation for writing this book was to prove the naysayers wrong. It isn't impossible to market fiction. You just have to be extra imaginative to market fiction.

You fiction writers should have expected that. Really.

Fiction authors who possess the imagination to invent entirely new worlds need to put that same creativity to use when marketing. The problem with marketing fiction, in my opinion, is that writers seem to lose all originality when placed in a position to promote their work. Shiny bookmarks and blogging about writing are not creative.

One of the goals of this book is to help you be genuinely creative. Too many marketing books can lull you into the dangerous waters of relying on "gimmick." That doesn't work in social media. We can smell a publicity stunt a mile away and we resent phonies. We will discuss some techniques for marketing fiction later, but first, let's go over some marketing basics. This goes back into understanding the *whys* behind your actions. When it comes to marketing your book (particularly fiction), what is the big goal?

Glad you asked.

In the world of social media there are 2 kinds of products. **Low-consideration purchases** and **high-consideration purchases**.

**Low consideration purchases** are of low social influence. Toothpaste. If I drop three bucks and hate the taste, it is no big deal to toss it and buy another...unless you are my mother. I am probably not very likely to get on-line and research the latest breakthroughs in fluoride so I can be certain I will be spending my $3 wisely. I will probably not go to blogs about whether Crest or Aquafresh is a better deal overall, and I can almost guarantee I will not need the support and approval of my peers when it comes to my choice in toothpaste.

**High consideration purchases**, however, are like cars, vacations, flat-screen TVs. These are products that peer pressure weighs heavily upon the decision. If I am about to drop a few grand on cruise vacation to Europe, I am far more likely to surf the web for comments and articles and blogs and any possible complaints. If I am going to buy a car or a riding mower, I will look to magazines and consumer reports to get the best product.

This type of purchase is heavily influenced by social definition and peer pressure. Sometimes facts are secondary to emotion and what that purchase comes to mean in a social context. Mac computers are a good example. Their advertising made Mac a social definition, with Mac computers being synonymous with being young and hip and cool and PC being the contiguous butt of the joke. It made no difference to Mac that they were comparing Apples and oranges, no pun intended. Even the least expensive Mac continues to be almost double the price of a PC. To place these two "competing" products in another market, that would be like Porsche comparing its Carrera against the Ford Mustang. Yet, whether you are a PC fan or a Mac fan, you have to admit their marketing worked.

How many consumers ran out and bought Mac computers for the image it portrayed? Many couldn't tell you two accurate facts about the size of the hard drive or the operating system. All they knew is they looked cool when working at a small table in their local Starbucks. Heck, I would be lying if I said that I wasn't tempted to buy a Mac laptop simply because it looked snazzier than my HP. The high-dollar purchase now has become more about the consumer's definition within society than about the product itself.

## Ah, but what about books?

But what about books? Where do they fit in? Some books don't cost much more than the tube of toothpaste, and they certainly don't cost nearly as much as a computer. So are books low or high consideration purchases?

Good question. Before we can discuss whether or not books are low or high-consideration purchases, I need you to do me a favor first. I want you to shrug off the *mythos* of the voracious reader who devours paperbacks like they are candy. Do you want these types of readers to be your fans and loyal to you as an author? WOW! Yes! But people who have a book habit like this are a totally different animal. Books, to them, are a low-consideration purchase. These consumers love books and forego all other activities to read. They don't need peer review to guide their choices of what to read, and often will finish books even when they don't like them (my dad). These types of readers don't need to be dragged away from other competing hobbies like video games or sports. These people *want* to read.

But these readers are not most people, and I feel this is a huge stumbling block for writers who are trying to build a fan base. Authors (especially fiction authors) get so focused on this white stag, that they forgo a lot of promising opportunities at the expense of an anomaly. There are **a lot of brown deer** to bring home to the tribe. Wait too long on a myth and you will starve.

Reality check...

A very large percentage of Americans do not consider themselves readers. Of the percentage that does occasionally buy books, I guarantee you most of them buy far more books than they ever read. These consumers are most likely to buy novels on impulse while wandering a Barnes & Noble waiting for their kids to hunt down the last copy of *Lord of the Flies* or while standing in line at a drug store as the lady in front of them argues over an expired .50 cent coupon.

Unless some outside peer factor steps in, an author can be at the mercy of chance. Is your book placed well at a bookstore, drug store or airport? Sadly that type of placement is often given to the big-name authors. So if you aren't a big-name author, what can you do?

### The Presto-Change-o

I feel the wise author sets a strategy in place that is designed to make converts. Feel free to tap into reader circles and reader reviews and groups, etc., but these groups represent a very miniscule percentage of the ***overall literate population in need of informing or entertaining***.

Is it great to secure the undying loyalty of the book-a-week reader? Sure. But do you need to? How many books can you write? My opinion? Who cares if a consumer only buys one or two books a year if they happen to be your books? The regular person who reads only a handful of books makes up the largest part of the population. This person, if convinced by peers that a particular book is THAT good, can be some of the most loyal fans. Don't believe me? Two words. *Harry Potter.*

Ahhhh, but here is where the *presto, change-o* happens. Now an item that in one group (the avid reader) was considered a low-consideration purchase has now changed to a high-consideration purchase for the individual who rarely reads and for the most part doesn't believe he enjoys reading as a past-time.

Thus, when it comes to this type of consumer buying a book, peer review now suddenly becomes ***critical***. The largest part of the population is only going to read a book or two a year, if that. They are not likely to go to bookstores, or be members of book clubs, or read guest blogs or sign up for author newsletters. Yet, how much fiction marketing is taking place only in the areas where we are likely to find self-professed readers? What about the rest of society who makes New Year's resolutions every January to read more?

The largest part of the population is watching *Lost*, and hanging out at Starbucks and goofing off on Facebook. It becomes our job to convince these *potential fans* that our book is worth forgoing all other types of fun. To this massive segment of the population, books are a **high-consideration purchase.** We have to convince them to part with their precious time.

Always keep in mind that books are different than most entertainment media. Novels aren't movies or music (duh). Seriously, though. Even a long movie is 3-4 hours and we can do other things like fold laundry while watching a movie or while listening to music. Books, however, pretty much take our undivided attention for hours and days and weeks, depending on the book and how fast a reader we happen to be. In YA, a writer has a tough sell. You are competing against sports and texting and hormones and YouTube and on and on and on. In romance, you are up against jobs, kids, marriage, laundry, *Dancing with the Stars*, etc. Can the challenge be daunting? Sure.

But let's look at the bright side. The general unmotivated population is like a HUGE boulder sitting at the top of a mountain. A lot of potential energy. Get that sucker moving and there will be no slowing it down.

*Twilight.*

One of the reasons I think J.K. Rowlings and Stephanie Meyer have been so successful is that they were able to mobilize huge populations of people who did not define themselves as "readers." Like Mac computers, these authors harnessed peer pressure to make their books a social definition. Whether they did so intentionally or by accident is irrelevant. They did it. There are plenty of fans who bought every single hardcopy of the *Harry Potter* series, but didn't consider themselves "readers," and one would be hard pressed to find a teenage girl who hadn't read the *Twilight* series. Many teenage *boys* have even read the series, and yet this is the demographic that typically reads the least.

Peer pressure is invaluable for swaying previously untapped markets into the die-hard fan category. Get enough people in a certain peer group to read your novel or book and give a thumbs-up recommendation that *your* blog, articles, book, novel, etc. are worth the time, and you are already on the downhill run.

So forget about the avid reader. Well, don't forget so much as refocus. We want readers of all kinds to love your book, thus we need to learn to think like a potential fan and understand what motivates people to make a purchase.

This is just one more reason why authors need to plug in socially. There are a lot of "things" clambering to take up your reader—your customer's TIME. Why not your book? Social media at least affords you the opportunity to toss your hat in the ring. But how do you stand out? You aren't the only writer vying for the reader's attention. Here is where understanding the *why* becomes pivotal.

Before we discuss marketing, I want to make a point. Social media will capitalize on what is known as *relationship sales.* People will generally buy your book not because they are being pitched to and hounded, but because they "know" you and it makes them *feel good* to support who they know. I always recommend we keep the spirit of service first, but I also feel that a basic understanding of marketing will be of great benefit once you have to prepare content and interact.

Marketing capitalizes on three main things. **The power of emotion. The power of repetition. The power of peer pressure.**

## The Power of Emotion

Emotions are powerful things, and those in the selling business capitalize on your emotions. It isn't a nefarious plan of mind-control. It's just that companies understand humans aren't robots. They know we are emotional creatures. Thus, companies know that if they hope to get us to pay attention long enough to part with our money, they must speak our language—feelings. Think about commercials. Commercials tap into our primitive thinking centers in hopes of making us associate their name and product/service with some pleasant emotion. We are barraged with these tactics on such a continuous basis, that they have become barely even noticeable.

**Allstate**—"You're in good hands." Emotional. Getting in an accident is stressful. We like that there is a promise we will be taken care of in a time of need.

**Geico**—"So easy a caveman can do it." No one likes feeling stupid. Complex insurance makes us feel stupid. Geico claims to make insurance simple. Emotional.

**Coca Cola**—"I'd like to buy the world a Coke." Friendship. We all like friendship. Makes us feel warm and fuzzy.

**Michelin**—"Because we know what's riding on your tires (picture of a baby)." Safety/Family.

**Auto Zone**—"Because Do-It-Yourself doesn't mean you have to do it alone." Mentor/Security. We all have taken something apart and then had to hire a professional to put it back together. This ad helps us feel empowered (to do it ourselves), but then also gives an added promise that someone will be there for us if we end up in a bind.

Good marketing appreciates the power of our human emotions and seeks to drive purchases out of the thinking centers and into the emotional centers.

### Thinking Center

*If I buy this new Corvette on the spot, my wife will kill me. No, she will divorce me. No, she will just kill me.*

### Emotional Centers

*Gee. The sales guy is right. I deserve this. I've worked my whole life and always bought a practical car. The power of that engine did feel amazing, and I loved all the attention I got when I pulled up in this baby. Sure was better than driving the Honda.*

**Thinking Center**

*Another $60 a month for a security system isn't in the budget.*

**Emotional Centers**

*This guy is telling me that crime is on the rise. Up 4% in my area just last month. I could be next. What kind of mother leaves her children in danger if a mere $60 a month can prevent disaster?*

Sales is actually pretty simple. You want to know the secret to sales? It is all in this little pyramid.

Maslow's Hierarchy of Needs

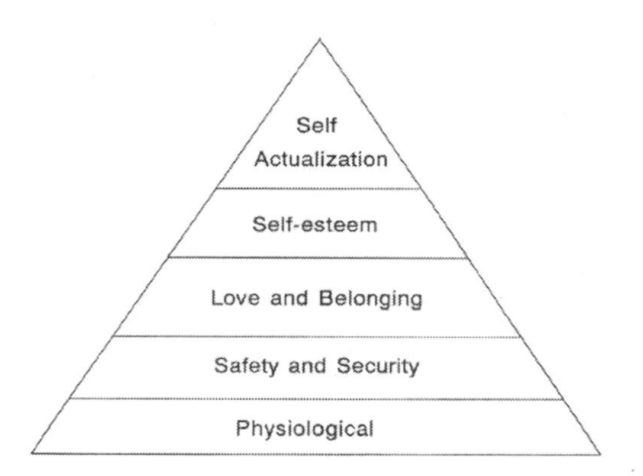


Whether we like it or not, marketing uses our human tendency to make decisions based on emotions. Humans are emotional creatures, and one is wise take that into account when creating any marketing plan. Applying emotional levers will be an important skill later when you start putting together an image and creating a platform.

Please get beyond some notion that we are in the business of manipulating people. Do you like being manipulated? I don't.

Good marketing doesn't manipulate people's emotions. It merely makes them pay more attention to their emotions and needs.

Think of it this way. Have you ever listened to a symphony? Hundreds of different instruments play together and often at the same time. Now, have you ever just focused on hearing one instrument? Like the oboe or the trumpet? This is what good marketing does. It interrupts the chorus of our lives and focuses us on one emotion or one need.

When you are writing your blogs (later) or even announcing your blog, you must harness the power of emotion if you desire to connect with others in a positive way. We will discuss more about blogs and content later, but which blog announcement is more appealing?

1) *Come read my new Warrior Writer Blog! "Writers—Do You Have the Right Stuff?"*
2) *According to the BEA, 9 out of 10 first-time novelists fail. How can you hope to beat the odds? Let me show you how.*

Go back up to Maslow's Hierarchy. Most people bee-bopping along on Twitter were just fine and hunky-dory working on their novel. They took a quick break to look at what was up in Twitter Land.

The first update isn't all that engaging. It shouts "Look at me! Look at me!" It does nothing to shake the viewer out of his comfort zone other than the small challenge of wondering if he has the Right Stuff (notice my title was emotive). This person might or might not stop by.

The second update, however, uses Maslow's Hierarchy. NF folk are all about filling a tangible need, so we need to rattle people out of the comfort zone. Comfortable people do not have needs for us to fill. This second update, however, is upsetting.

There is some guy out there who is all happy working on his first novel. When he sees this statistic, it is a punch to the gut. My announcement has now bumped him out of his happy place. He wants to feel better, though and that is where I offer a solution. My solution? I have posted a link to a blog that will help him beat the odds and win. This announcement is geared to showing this person a need that he might not have known he had when he woke up that morning. He might not have fully appreciated that the odds were so stacked against him. When he read my announcement, he *felt* fear. Fear of failure. I wanted him to feel that good fear because I was pointing out a genuine threat to his publishing success. The smart writer knows what he is getting into and faces the tough reality so as to better plan and prepare. But, I also wanted him to feel relief because I really dig helping people. Now, our writer friend is back up in his happy place, and maybe even a little higher on the hierarchy because he saw a potential threat (previously unnoticed), but then was *empowered* to have a different fate. Not only is he empowered, but he is grateful to me for being part of his team by pointing out the threat and then helping him conquer it.

Thus, when you are creating your blogs later, I really want you to always focus on what emotion you are bringing to the forefront, even if it is just wild curiosity.

Boring and self-centered—*Read my new blog! "The Disappearance of the Anasazi"*

Interesting and reader-centered—*Can science explain how thousands of people vanished with no trace? Click this link and find out.*

Oooh! Come on! You know we are just going to read this because we're curious. It's the only logical reason why six seasons of ghost hunting shows can not only survive, but prosper. They never find a single ghost, and yet we watch, transfixed, as cameramen traipse through basements commenting about how they "sense a presence." It is the only possible explanation why we sane, rational human beings would watch a show for an hour in hopes of seeing real proof of Atlantis or vampires or Big Foot. This stuff works! We fall for it every time. And if you blog well enough, people won't care that you didn't actually have the real proof that Nessie exists or that space aliens have taken over Congress. You hooked us by intriguing our curiosity instead of being lazy and spamming us.

We were entertained, so we forgive you and we like you because you are interesting.

For fiction authors. You guys fill a need just like the NF folk—a need to be inspired and entertained. When you go to pitch your book using tweets and status updates, I want you to think **movie log line.** Emotional.

*What if the world's fate lay in the hands of one vampire? Find out here*

*Does true love ever die?*

*Can one person change the world?*

All right. Lots of fun, but we must move on. The power of emotion becomes even stronger when coupled with…

**The Power of Repetition**

Back when I was in sales, we used to say, "Say it once. Say it twice. Say it three times. Say it four times. Say it five times and they will believe." And this sounds like some reprehensible brainwashing technique, but it really is just another reality of the human condition.

Why do you think companies pay good money to air commercials over and over and over, especially when launching a brand new product? If virtually every time you turn on the TV, consumers hear how "*Such and Such Movie*" is blowing away all the critics, what is the first movie they are likely to think of once Friday date night arrives? Companies realize they have to power to influence our choices just simply by telling us something over and over and over until it is embedded in our minds as truth.

Repetition is powerful. Those in marketing understand this. This is why, when we discuss branding later, it is absolutely **crucial** for you to brand your name. If people see your name over and over and over and it is always associated with your content, that is like a non-stop commercial pitching your work every single day. This is why a moniker can absolutely KILL your platform.

**When you use anything other than the name that will be printed across your book, you give up your most valuable marketing real estate...the top of mind.** Every time you "tweet" or send out a status update, you want those following you to see your name. It is like your very own commercial playing over and over and over, scrolling down the news feed.

***This goes beyond posting "Buy my book! Read my blog! Me! Me! Me!" Then you just become about as engaging as a pop-up ad.**

If Maura Devlin (made up name) regularly posts blogs on fantasy and links to other fantasy events and talks about her latest fantasy novel that will soon be released, guess what? When I run by a bookstore, I will default to what I know...and now I KNOW Maura because I have basically had scrolling "commercials" from her every day I am on Twitter.

I also feel like I am Maura's virtual friend, and I like to support my friends first. So if I am going to try something new in fantasy beyond staples like J.R.R. Tolkien, Piers Anthony, or Anne McCaffrey, I am going to try Maura Devlin because she has focused all her social media energy to making her name synonymous with good fantasy entertainment.

Let's use Maura as an example.

**Scenario 1, Maura is Dragon_Girl**

On Twitter, I see a lot of:

@Dragon_Girl New "Wizard Woman" blog post. Where did the legend of dragons begin? (inserts link here)

@Dragon_Girl Book coming out soon. Should be here by end of May

@Dragon_Girl I love the cover. What do you think? (She attaches the cover here)

@Dragon_Girl Book signing is this weekend. Make sure you are early before we run out of books (attaches information on how to get to book signing)

***Notice I NEVER see Dragon_Girl's NAME. She is always top of mind, but using the WRONG NAME. Even if I wanted to buy her book, I would be at a loss and would have to go do research. If I have an antsy husband who wants me to hurry and get my book so we can go to Costco, and a baby who is teething and starting to fuss, I am not that motivated to figure out Dragon_Girl's real identity.

**Scenario 2, Maura Using Pen Name Maura Devlin**
On the contrary, I SHOULD see a lot of:

@Maura_Devlin New "Wizard Woman" blog post. Where did the legend of dragons begin? (inserts link here)

@Maura_Devlin The dragons are near! Book coming out soon. End of May!

@Maura_Devlin I love the cover. What do you think? (She attaches the cover w/dragon art here)

@Maura_Devlin Book signing is this weekend. Make sure you are early before we run out of books (attaches information on how to get to book signing)

Maura Devlin doesn't need to be Dragon_Girl for those who follow to **get** that she writes fantasy. We are actually pretty sharp. This second scenario keeps Maura's name continually top of mind so that those in her network see a scrolling stream of, "Maura Devlin, Maura Devlin, Maura Devlin...always linked with her content—dragons/fantasy."

Repetition is powerful, so make sure you are using it to your advantage. Additionally, if you are going to repeat something, make sure it is always framed as a positive. In marketing nothing is ever said accidentally. Words have a particular order and use a certain verb for a very good and highly strategized reason. This is why, you too, also have to be very careful what you say and how you say it.

Did you know the mind has this really odd tendency of chopping off conditionals, and only beginning to listen at the first active verb? Here you thought all this time that your kids were just being defiant. I am dead serious. That is one of the reasons it is counter-productive to send out "negative goal messages."

For instance:

*Don't forget to buy my latest book.*

If the human brain picks up on the first active verb, what message is being heard?

***Forget** to buy my latest book.*

Rather, this author should post positive goals, like:

***Remember** to buy my latest book.*

Now, with a little sentence revision, the subconscious cue you are giving is a positive one with a greater likelihood of having a positive outcome. If you are going to repeat something over and over, make sure you use active verbs and focus on the positive. You will get a better result for all of your effort if people see this 20 times:

*Make sure you sign up for my writing workshop today!*

*Be one of the first to own my newest book,* Dragon's Keep. *Pre-order today and save 15%!*

*Remember Mother's Day is two weeks away. Save time and give Mom a gift she'll love, a copy of my latest devotional.*

I know this seems very silly, but it does work. The subconscious mind is an amazing thing, and if we can harness it to our favor, we can achieve almost anything.

I recently challenged a group of agents to start talking about the writer behaviors they would like to see instead of giving a long laundry list of pet peeves. When a group of nervous writers hear, "Don't panic at your pitch session. It isn't the end of the world," what are they really hearing? *"Panic at your pitch session. End of the world."* These agents would have gotten far better results if they framed this in an active, positive statement like, "Stay calm at the pitch session. We are here to help you."

Make using positive statements a habit, not only on social media, but in life. You will be astounded at the results.

Finally when the power of emotion meets the power of repetition, it combines to form...

## The Power of Peer Pressure

Peer pressure melds the power of emotion with the power of repetition to create the most powerful buying force. Suddenly teenagers are wearing pants ten sizes too big and their shirts on inside-out. Women are willing to spend a half a paycheck on a purse, and men think a Dallas Cowboys decal on the side of the family mini-van looks cool.

Peer pressure can drive normal people to make some very irrational choices and feel good about those choices. The pressure for humans to not only conform but to also be admired within a social group is probably the most powerful motivator (back to Mac and PC). It is rare to find individuals who are willing to go against the collective. Add an "expert" opinion and peer pressure is a virtually irresistible force. Well, 4 out of 5 will agree.

The plain truth is that we often rely on our peers to form and validate our choices. If birds of a feather flock together, then it is pretty safe to assume that our groups are comprised of people who are a lot like us. They have similar tastes, likes, dislikes, standards, and values. We trust them because they are a reflection of who we are and often what we aspire to become.

This is where the power of networking becomes invaluable, especially when it comes to this newly emerging ecommerce marketplace. If you tap into emotion and make the person feel empowered, he likely will pass on your message (repetition). If enough people pass on your message, suddenly you have mobilized an entire peer group to your favor (peer pressure). Peer pressure is not always a negative thing. Our peers make us try harder and reach for more. If it weren't for peer pressure, we wouldn't have as much of a desire to be thin and fit, or work extra hard to be successful. We wouldn't get out of bed on a Saturday and wear pink in the freezing cold to fight breast cancer, or willingly pay $10 for an eco-friendly light bulb instead of a buck for a regular light bulb. Peer pressure, when harnessed in a positive way, can change the world.

Positive peer pressure is a writer's best chance to marshal that massive group of people who do not consider themselves readers to eventually become die-hard fans.

*We understand, Kristen! But how do we do this? You are killing us!*

Okay. Okay. You guys have been patient. Ready to get started?

## Act II—(Technology Section) Building Your Social Media Platform

Welcome to the future!

Right now I know all of you are just chomping at the bit to go and build your social media platform. Hold on! Part of why you are reading this book is because you desire to stand apart from the crowd and be noticed. Right? I am your trusted guide and my goal is to help you do just that. The wise writer employs strategy ahead of time. Remember, you have brilliant, best-selling books to write. Social media can either be fun, productive, and fulfilling, or it can be a tedious chore that just makes us hate our computers.

This book is designed to give you a distinctive advantage over your competition. How? You are going to have a solid strategy in place before you ever begin. This technology part of the book will be in three stages. *Stage One* will guide you through understanding and gathering your content. *Stage Two* is where we will actually open accounts and build social media pages. You will be building a Wordpress blog and then pages on all three major sites. The goal is to command as much digital real estate as possible. This might feel a tad overwhelming, but we are going to do this together, and it really is pretty simple. *Stage Three* will be what is called optimization. By the end of these three stages, you will have a solid social media foundation designed to connect with readers. Also, and this is the great part, you will be able to maintain and grow this platform easily. My method will help you have a solid presence on all the platforms, without having to give up huge chunks of time.

## **Stage One—Understanding & Gathering Content**

This section is going to help you understand what makes good content. Too many people get all excited and rush out and sign up for Facebook and Twitter, but then have nothing to say. So often what ends up happening is that their poor little Twitter account will be abandoned with three tweets and two friends. Or, since content wasn't prepared ahead of time, people will drift along chit-chatting with friends and playing Mafia Wars and will have lots and lots of activity, and little to no productivity.

### **Your Brand**

If you are a writer desiring to use social media to build a platform, you have one huge goal. Become a brand. If you build your platform properly, you will be on your way to making *your name synonymous with your (entertaining, interesting, informative) content.* THAT is the goal. Stephen King will always be tied to horror, and Danielle Steele is tethered to romance.

**Being published is not the real end goal. Being published is only the means to your real end goal—SELLING BOOKS.**

Kristen! Must you be so crass?
Yep.

Plain truth is this. Great, you get published. But, if you don't sell enough books, you cannot quit your day job. If you fail to sell out your print run, you hurt your chances of another book contract. In order to do what you love–WRITE–you must learn to do what you hate–SELL. It doesn't have to be as hard as a lot of people make it. Brand your name, then your name can do the selling while you do the writing.

In order to maximize sales, your goal is to become a brand. Brand=Big Sales

If I want a good thriller, I pick up a James Rollins. If I want a good YA book, I pick up Stephanie Meyer. A good legal suspense, read John Grisham. Amy Tan will have to change her name if she decides to suddenly start writing novels about the Italian Mob. These authors are the designer brands of writing.

Most of us don't have time to research each and every purchasing decision and thus, we as consumers, are prone to rely heavily on brands. Brands let us know what to expect. When we see Levis, we expect clothing, usually of the denim variety. We expect not only a certain type of product but, over time, we come to anticipate a certain kind of quality. Honda is a brand. When we buy a Honda, we in turn expect a great car that is gas efficient and reliable. Nike is a brand for a good running shoe. So on and so forth.

As a writer, your goal is the same. Your big goal should be to link *your name* interminably with *your content.* Produce enough good content and eventually readers won't need to read every review about your latest book before they buy. They will trust you for good product and will pre-order your books because they have confidence you provide content that is entertaining, interesting, or informative.

Thus, before you sign up for any social media site, the very first thing you need to decide is your brand name. Before we go any further, we need to have a serious discussion. Branding is the single largest stumbling block I see for writers who desire to use social media in order to build a platform. The **absolute only acceptable username (brand) is the name you desire to publish under**. Period. This is a crucial step and enormously essential you do this one thing right in order to 1) be effective 2) be able to link all of your platforms together to make social media simple and manageable 3) begin building a solid platform.

The Internet has valuable real estate that you will want to command. How you claim that digital real estate is by using you **name**. If you use anything other than your name, you render most of the tools I give you minimally effective.

Writers' loooove being creative with their usernames. I know. I didn't always do everything perfectly and learned a lot through trial and error. Yet, one day I realized there was good news and bad news. The good news? Hundreds, even thousands of people knew I was a writer. Great! The bad news? They knew texaswriterchik was a writer, NOT Kristen Lamb. A potential reader could not go into a Barnes & Noble and request the latest book by texaswriterchik, and Amazon likely wouldn't know texaswriterchik either. And it was insanely egotistical for me to expect that a reader would actually take the time to go hunt down my real name when they could choose from all these other authors who'd been smart enough to actually brand their name.

That cost me a YEAR rebranding "Kristen Lamb." It meant nothing that texaswriterchik was always top of mind to my followers with great content. I had failed to brand the most essential component—me. Not only that, but by the time I tried to get "Kristen Lamb" as a username, it was taken. Yes, there was more than one Kristen Lamb. In all likelihood your name is probably taken as well. Here is a good place to get creative. I became "KristenLambTX." Most people are just going to focus on your name, so what comes after is of little distraction. I could have also done KristenLamb_author, KLamb, the_KristenLamb or KristenLamb007. You get the idea. Kristen Lamb still is the focus for anyone following and it keeps my name, *my brand*, top of mind.

Normally, I like to teach by telling you **what to do, as opposed to what not to do**. But here we are going to take a break from that approach, and for very good reasons. There are social media classes that will give tips and tools and techniques that are fantastic for the corporate world, but they can KILL a writer's platform. These gurus will make some of these techniques sound fantastic and very appealing, because in another world these tools do really well. I have even tried some of them in the past, so I know why these techniques don't work for writers.

Part of why I decided to write this book is because I have made all the mistakes. I know what seems like a good idea, but in application is just a proverbial tar baby. So think of it this way: **I have wasted all your time for you.**

It is too easy to be lulled down a dangerous and ineffective path. Thus, before we go any further figuring out your brand, let's look at some of the largest mistakes I see authors make, and why they are unproductive techniques when it comes to the world of publishing.

### Mistake #1—Branding the Title of Your Book

I have seen a lot of faulty instruction when it comes to creating a brand. This is one of the areas where I feel tips used for Corporate America break down when it comes to publishing. If I open a restaurant and I decide to promote that restaurant through social media, I know a) I control the name b) I also know the odds are likely the restaurant's name will remain the same c) that short of selling or going out of business, I will need to promote that restaurant until I retire or decide to get out of the restaurant business. Same with a boutique or any kind of small business.

Yet, I have seen too many writers try to take a similar approach by using the name of their book. HUGE, HUGE, HUGE MISTAKE! There are a number of reasons why this is a critical error.

#### Unless you self-publish, you have no control over title.

Say our friend Jack Spywell is writing his first thriller, and has named this opus *Merchant of Vengeance*. Great title. Jack spends the two years it takes to research and write this thriller also building a huge social media following. Smart move, which I highly recommend. *Merchant of Vengeance* is on Facebook and MySpace and has fan pages and blogs. Jack has worked his tail off to have readers eagerly awaiting his novel. Jack gets an agent who loves his work and a publishing house ready to offer a deal. Only one problem. Jim Super-Mega-Thriller-Best-Selling-Author is coming out with his latest book called *Merchant of Menace*. No worries, though. The editor loves Jack's book and thinks it will do great...but the title needs to be changed.

Ack!

Yeah, our friend just spent two years branding a name that isn't even going to be used. To make matters worse, Jack's followers only peripherally know Jack Spywell, since *Merchant of Vengeance* has been top of mind for two years. Not only will Jack have to rebuild his entire platform, he has cost himself two valuable years he could have been branding *Jack Spywell* as THE author for thrillers.

**Business owners can control the name of their businesses. Writers have no control over the title of their books.** Usually a title will be changed for business reasons like the example above. The publishing house is on your side. They want you to sell lots and lots of books. They make money if you do. Thus, it makes no sense to spend time and effort branding something that could change in an instant and collapse your months or years of hard work.

**If you write more than one book, you have to start all over again.**

Even if our pal Jack had gotten his way and been able to maintain the title *Merchant of Vengeance*, if he writes more than one book, he is back at square one. How effective is Jack going to be if every time he has a new book he has to build an entirely separate platform? And to make matters worse, Jack is likely to frustrate and lose followers along the way.

**Too many platforms render any author ineffective at his craft.**

Writers need time to produce the product...books. Unlike a social media platform for a personal trainer or a florist, an author's social media is difficult to outsource to a firm. Fans/readers will look to the author for a majority of the content, unlike a café owner who can hire an intern to regularly post coupons and promotional items. Followers do not expect a two-way dialogue with Domino's Pizza. They will expect this kind of dialogue from authors (until you are so huge that fans forgive you). The best social media platform is one that is easy to maintain. That is why it is crucial to brand the right name.

### Mistake #2—Branding Your Content

I have had quite a few intense discussions about this. You are the brand. Your name is your brand, not what you do. So, say our imaginary friend Amy Readmuch writes about the History of the Roman Empire. Using Roman_History_Gal is simply wasting precious time she needs to establish her as an expert regarding this topic. Remember my blunder with texaswriterchik? Same thing. Amy's name is most important. Now, Amy does need to have a complete brand—***Amy Readmuch Roman History Author***—but her name *must* be part of the brand, because a potential reader cannot find Roman_History_Gal or SPQR_Chik at the local bookstore.

If people follow Amy's content, and she is doing what she should, it won't take long for others to figure out that their pal Amy is all about the Roman Empire.

### Mistake #3—Branding the Name of Characters

The best approach, again, is to build a platform for *you* using *your name*. In order to avoid redundancy, I will just say that branding a character is a bad idea mainly for the same reasons it is a bad idea to brand the title of a book. An agent or editor might want your book, but only if you do some revisions. Those revisions could mean name changes for major characters or even the removal of a major character. You don't know and it again is something you cannot control.

Also, most new authors dream of having a series with the same characters. Well, if the first book flops, what are the odds of a series? Or, maybe by book two, you decide you want to write something different. Then you are back to the same issue of having to start all over for new characters.

Pen name is something static that you can control, thus the wisest choice for brand.

Lee Child is probably best known for his Jack Reacher character, but *Lee Child* is the brand, not Jack Reacher. Also, this gives Child the option down the line of writing a book that is not a Jack Reacher book.

### Mistake #4—Branding Multiple Identities

When I teach my workshops, I often ask two silly questions. First, "How many of you do more than one thing?" Usually everyone raises their hands. Second, "How many of you *know people* who do more than one thing?" Again, lots of hands.

**If you do a number of different things (i.e. you write fiction and non-fiction and teach workshops), it is almost always unnecessary to have multiple identities.**

Here is the simple truth. It will not rip the fabric of my reality to know that you do more than one thing. Thus, a wider brand is a good idea if you do a lot of things. Think *Proctor & Gamble.* Most American consumers know *Proctor & Gamble* is the banner that flies over a broad range of consumer products worldwide, from personal care to food products. Pampers, Tide, Bounty, Folgers, Iams, and Clairol are just a few of the brands beneath the major brand.

Most big brands have more than one "product" they represent. *Calvin Klein* is the brand that includes purses, shoes, jeans, kid's clothing, underwear, teddy bears, bedding, and even perfume. We are okay with that because all of the "products" are at least tangentially related. Now if *Calvin Klein* suddenly decided to branch out and include designer auto parts, that would certainly throw us for a loop. But, so long as *Calvin Klein* doesn't get weird on us, we manage to keep up just fine.

When creating your own brand, my advice is to keep your brand consistent. For instance, my brand is Kristen Lamb. Guess what? It has yet to upset the space time continuum for me to announce on Twitter or MySpace that I write fiction and non-fiction. I teach novel-writing workshops and social media for writers. I blog about writing. Now if all the sudden I tried to graft landscape design into my brand (no pun intended), I would probably confuse a lot of followers. But, so long as I maintain a consistency—writer, writing, fiction, non-fiction, all things literary—I am just peachy. People are smart. They will *get* that you wear a lot of hats. You really do not need separate social media identities for everything. Most of the time, all that tactic does is spread you far too thinly and it dilutes how well your followers/fans can support you.

I can put all of my "products" under one umbrella of *Kristen Lamb* to make things easy and simple for my fans. Through this "brand" I can then direct traffic. For non-fiction social media brilliance, go here to *Kristen Lamb.* But for Kristen's wildly exciting romance novels, go to *Kristen Lamb **writing as** FiFi Fakename. Fifi Fakename* should be a separate page on my main *Kristen Lamb* website to mentally cue the guest browsing my site to keep *Kristen Lamb* as *NF Author* and *FiFi Fakename Romance Author* separate. I could have a totally different website, but why? One click is one click, whether that it to a separate page on a singular web site or to a totally different site altogether the effect is the same.

Make no mistake. I will have to work twice as hard. I will have to maintain and promote ***Kristen Lamb NF Author*** and ***Fifi Fakename Romance Author*** with equal fervor. That is the downside of multiple identities and one of the reasons it is easier to pick one thing and be really, really good at it. If you wear multiple hats, you will have to work harder, but my point in this section is that this is already hard. No need to make it even harder.

Now if my romance writing takes off and is incredibly successful, then I might consider giving my *FiFi* identity her own web site and begin creating *Fifi's* social media identity. If you run into the same circumstance, feel free to separate off your identities and expect **a lot** of work (good time to hire my friend Fred and outsource). But most of us aren't to that point yet. My goal is to help you create a social media platform that is linked together, self-sustaining, and manageable. If you have more personalities than Sybil running around unattended, then you are just making too much work for yourself. And please trust me when I say I learned this the hard way.

One big name brand works just fine in the world of writing. Most people who follow NY Times Best–Selling Author Bob Mayer *know* that he writes science fiction as Robert Doherty. He doesn't need an entirely new web site and Twitter identity to keep from short-circuiting our brains. We get it.

Balancing multiple "identities" is tough and a lot of work and there are ways to manage them, but we won't discuss them in this book. But for now, suffice to say that one big brand name is what you need to focus on with all your effort. Most of us will have our hands full with just that. We do have day jobs and families and need to sleep.

****Note: If you are doing so many things that you need 10 different identities, it might be a good idea to take an inventory and do all you can to get focused.*

Just because Twitter allows you to have multiple identities doesn't mean it is a good idea, **especially if you are unpublished**. If you are building a social media platform in an effort to make you/your work more attractive to and editor/agent (which is a great plan), then I advise you think of things from their perspective for a moment. Which is more appealing?

Writer A has created pages on the three major platforms under ONE name (the pen name) and linked those domains together. Collectively, Writer A has thousands of followers who all know her *by name* for her **content (**fantasy fiction) and follow her regularly posted blog about fantasy.

Writer B also has pages on all three of the major platforms. Lots of pages and lots of identities, including fan pages and five different Twitter identities, one for each of the dragons featured in her books. Writer B also blogs from each of her character's viewpoints. Clever, but how effective?

One says, "Strategy," and the other screams "Gimmick!"

Which writer looks like a better investment? Which writer appears more focused? Which writer looks more professional? Which writer looks like she might actually have time left over to write more books?

Now you get the idea.

Okay, enough of that. Now we are ready to move forward. Write at the top of a blank piece of paper the name you wish to brand.

## Goals

Like most anything else in life, it helps to set clearly defined, manageable and achievable goals. It is the simplest way for you to measure your progress. Why is that important? Because I want you to always be encouraged that you are enjoying forward momentum. Also, if you happen to hit a snag in your plan, you will spot it far sooner and be able to adjust your approach. Your best bet with social media is to maximize your impact. That might seem like a no-brainer, but I have been witness to far too many writers who spent months or even years building the wrong platform and branding the wrong name. But since you know better and already have your brand name, now we need to get an idea of what you desire to accomplish.

Not all writers are at the same stage in the game. If you are new and haven't yet started on your novel and don't really know even what genre you desire to write, it is still a good time to begin building your brand. Your brand will just be very broad in the beginning. *Your Name Writer*, for example. Over time, you should be able to eventually narrow your brand and bring it into more focus. But if you know that you always wanted to write spy novels, then go narrow and do it well. *Your Name Spy Novel Writer.*

At such an early stage in your career, your social media goals will lean more heavily to networking with others in your industry. A new author needs to learn more about the craft and the world of publishing than, say a seasoned, multi-published author who is using social media to communicate with fans. A published author doesn't need to learn how to write a query letter. Smart green peas do. By the same token, a published author needs to network with people who can teach her how to drive up book sales. An unpublished author shouldn't focus on that information if he hasn't even finished his first novel.

Thus, take a few minutes to set some goals. In my novel writing workshop, Warrior Writer Boot Camp, this is the very first step for all attendees so as to keep them focused and accountable.

**You have one Strategic Career Goal as a writer**. The easiest way to figure this out is to answer this question: Where do I want to be in five years as an author? A Strategic Career Goal is important because it sets the creative and practical processes in motion. For instance, when I took Bob's Warrior Writer Workshop, I suddenly realized the crucial difference between, *"In 5 years I want to be a best-selling author,"* versus *"In 5 years I want to be a best-selling* ***thriller*** *author."* The first Strategic Career Goal was far too broad and would have made it difficult to plan. Best-selling author of what? Origami Cookbooks? Part of the earlier discussion about brand was to help you be able to forge that one overall Strategic Career Goal that would guide you in your writing career. I highly recommend Bob Mayer's *Warrior Writer—From Writer to Published Author.* He has a great template for goal-setting when it comes to your career as a whole. Regardless, though, the Strategic Career Goal will serve to keep you accountable and focus your social media marketing.

**You have one Strategic Marketing Goal for social media.** Your goal is to create a brand, an image that pops into people's minds when they see or hear your name. If you are a new author and haven't even written a book, your key goal will still be to create an image, even if that image is *Your Name Writer.* When I first started out, I had no clue what I wanted to write, so I went very broad and at least sought to establish that when people heard *Kristen Lamb*, they thought *writer.*

This one Strategic Marketing Goal determines the subordinate Tactical Goals. Put plainly. What are reasonable steps to take to achieve that one large goal?

If you want to become a brand name author and you haven't even learned how to write yet or figured out what genre you'd like to commit yourself to, then you will have a longer list of Tactical Goals. It might looks something like this for a newbie.

Strategic Career Goal—In 5 years I want to be a best-selling Regency Romance author.

Strategic Marketing Goal—I want *Fifi Fakename* to be synonymous with great *Regency Romance*

Tactical Goals:

1) Create a platform with a Regency Romance image.
2) Add 5 best-selling Regency Romance authors to my network per week on each platform.
3) Follow as many romance authors as cross my path. 25 per week min.
4) Follow all the publishing houses I would like to one day publish my work.
5) Follow agents who would be a future good fit for my manuscript.
6) Blog once a week to establish my authority as a Regency Romance author.
7) Network with other authors, published and unpublished.
8) Join groups that can help establish my reputation as a Regency Romance author.

9) Read every blog about the craft of writing posted by my top authors, agents and publishing houses.
10) Pay attention to the habits of my heroes and see if there is anything I can add to my own routine.

You don't have to be super-complicated, but goals are going to be very helpful to keep you focused. If you are an unpublished newbie, then goals should revolve a lot around learning and networking. If you are published, then you should also incorporate goals to widen your network and drive up book sales. I believe that once you have read this book, setting goals will be far easier because you will have an idea of the big picture. But, it would be a good idea to write down your Strategic Career Goal, your Strategic Marketing Goal and then ten Tactical Goals. Feel free to modify them later.

At this point you should have your brand name (your name), your Strategic Career Goal as a Writer and your Strategic Marketing Goal. The goals will be important in determining which social media platforms will be the best fit.

## Tags

Now, beneath your brand name, I want you to write nouns that describe who you are and what you do. These are called **tags**. **Tags** are keywords or terms, **metadata** (data about data) that make information searchable. Sounds complex, but really isn't. I'll explain more about this in a minute.

So write tags. Lots of them. The more tags you write the better. My list of tags looks like this:

Kristen Lamb—Kristen Lamb, author, writer, We Are Not Alone, speaker, teacher, fiction, non-fiction, social media, marketing, PR, public relations, social media for writers, editing, writing, writing workshops, platform, brand, publishing, books, novels, self-help, Twitter, MySpace, Facebook, blogs, blogging, blogger, Wordpress, web sites…

These words should describe you and your overall content. Think of your tags like ingredients that make up YOU as a brand.

As you can see from the above list, all of those nouns after my name go together. I didn't include X-Box 360, hiking, gourmet cooking or dogs. Those nouns would tell about Kristen Lamb as a person not Kristen Lamb as an author (brand).

I suggest you free-write as many tags as you can, then highlight your favorites. Give the list to people who know you and ask for their favorites. Sometimes others see us more clearly than we see ourselves, so enlisting the help of friends, family and colleagues can be invaluable. Ask them to add any tags that you might not have thought of.

This exercise might be silly, but it is a vital step toward helping you focus, and making you more visible on the Web. Search engines use these tags to find you and your content. Think of a search engine (Google, Bing, Yahoo) as your own personal genie who wants to give you exactly what you desire...or pretty darn close. Tags help the search engine find exactly what you are looking for. Thus, the more tags you have, the greater your odds of being selected on the first page of a Google search.

An example.

On Google, I type, "How do I use social media to create fiction brand?"

Google hones in on key words and sees this—"How do I use **social media** to create a **fiction brand**"—and then searches for anything tagged with those key words. A search engine will seek out content that is tagged FIRST. Only afterwards, will Google search for words in the body of the content.

Thus, when I blog about social media, my best bet is to include likely tags associated with my content (**social media, fiction, brand**). This will bump me up higher in position than if I write an article with these words, but then fail to place them in tags. My article might be better than another article, but if the author of the competing blog took the time to tag and I didn't, he will rank higher in the search, and I risk losing a reader. How many of us are motivated enough to go to the second page of a search? We aren't. We look to the top positions. Tagging can make the difference.

## Your Profile Picture

Most social media platforms will have a place for your picture. I strongly recommend a good headshot: clear, minimal background distractions, attractive, and brightly-colored or in some way eye-catching. Nowadays, computers make it easier than ever to create a great profile picture. Some people will use avatars and logos and pictures of their dogs. Those are all great options for the casual social media participant who is on-line for fun and games.

You are different.

**In order to build a platform, it is imperative you be viewed as an expert.** Experts usually use their pictures. You want to stand out as an authority on your subject, even if your subject happens to be elves and wizards. Headshots also serve another vital purpose. It's easier for others to feel attachment to a picture of your face than it is to feel fondly about your really cool logo. Your goal is for those in your social media platform to feel vested in you personally. Headshots subconsciously communicate to others that you are different than the casual user. If you still aren't sure, I encourage you look up the social media sites of the top authors in your genre. I guarantee they will have a picture of their face (usually the picture used on the backs of their books).

If you don't have a suitable picture, no problem. Dress up nicely in appropriate attire. Your wedding dress might not be the best choice (unless you write books about weddings). I recommend a bright, flattering color. Remember, your picture will likely have to be the eye-catching component that will make you stand out in a sea of other people. The human eye is drawn to bright and shiny objects.

For women, I recommend you put on about four times more makeup than you normally would wear, as if you are going to a dark nightclub. Seriously. This will make your features more distinctive in the photograph and offer better odds of a good headshot. Models, actors and news anchors all know what plastering on the cosmetics can do for the face. And, yes, you will look like a clown in person, but this will make a much better, clearer photo. If you aren't very good at doing your makeup, make an appointment for a makeover at your local department store's cosmetics counter. Many of the cosmetics companies (Mac Cosmetics is my favorite) have professional makeup artists who are usually bored silly during the week. They will spend lots of time with all the best techniques and products making you look red carpet ready. Most of the time the makeover is free with a minimal purchase.

To take this wonderful profile picture, stand against a simple, contrasting background. I happen to have light blonde hair. Thus, a light or white background is a bad idea. Dark backgrounds are better. If you are dark-headed, then opt for a lighter background. Make sure you to minimize any background clutter. You want the focus to be on you.

Take the picture with a good digital camera. Yes, I know phones and webcams can take pictures, but grainy unfocused images are distracting. Pay for a good headshot if you have to. It is a wise investment for any author.

Remember, one huge goal with social media is to create emotional attachment to you. It is much harder for others to feel personally vested in your latest book cover, a picture of a man's six-pack abs, your logo or your cat.

## Gather Your Content

Think of the image you desire to go along with your brand. **Content is anything you can post on the web to enhance and build your brand. Articles, photos, video, book reviews, you name it.** If it falls in with the image you want to create, put it in a file. If your content is not digital, you may need to scan it into your computer or retype it so you can use it on the Web. Pictures of you at writing conferences or in writing groups are great. Pictures of you kayaking? Well, if you write about aliens then maybe not as valuable as pictures of you at a Trekkie Convention.

Gather all the content that fits in your brand and then see how many of those tag words you can assign each piece of content.

Now, you may not have a lot of content, and that is okay. I suggest you write at least fifteen short articles (will be used for blogs) that have to do with your topic. I call these blogs **strategic blogs or strategic content.** Make them 500-800 words in length and something that would engage conversation by potential readers.

If you write fantasy, blogging about writing will only appeal to writers. The goal of your platform is to reach out to readers. So dig deeper. What type of fantasy do you write? What is the subject and how can you make it engaging?

Let's have some examples.

If you write J.R.R. Tolkien-style fantasy, you might have strategic blogs that look like these:

**J.R.R. Tolkien. Why are his books so appealing even after all these years?**

**Dragons—Where did the myth begin?**

**Why are fantasy fans so loyal?**

Do you see how these titles are geared to engage **fans/readers** of fantasy fiction? It gives the fantasy author an opportunity to show she can write and be interesting, but also serves to establish her as an expert in the area of fantasy (content). Blogs such as these also entice readers to engage and offer their insights and opinions and thereby spark discussions.

Your goal is to post on topics related to your content. If you are writing historical romance, blog about romance in that time period. How were women treated? Maybe write about arranged marriage and the practice of betrothal. Blog about societal attitudes of the time toward choosing a partner, marriage, children, the role of women, sex before marriage. Compare and contrast that with today's practices. Be engaging. Write about historical idiosyncrasies that would be interesting to readers who like historical romance of that time period.

If you write sci-fi, you could write about time travel or breaking the speed of light or wormholes. Blog about Stephen Hawking's theory about extraterrestrial life. Blog in ways that communicate your distinctive writing voice. If your novels tend to be funny, then be funny. If your novels are serious, then be serious. But, regardless the topic or tone, above all **be engaging.**

I advise you sit down and compile a list of possible blog topics. Shoot for 100. If you can't reach 50, then forget about it. Adjust your brand. Ask yourself if this is something that you could see yourself staying excited about writing for years and years. This is part of why I recommend getting creative with topics by profiling the reader. Then that frees you up to blog on more than UFOs. Now you can blog on what would interest people interested in UFOs. That is a lot more topics to write about. This diversity will keep you and your followers from being bored to tears.

Writing (blogging) about your content will do a number of things:

1) It will connect you with readers who like your subject area (ghosts, romance, Civil War, spies).
2) It will likely interest your fellow writers without alienating your readership.
3) Blogging on content establishes you as an expert in regard to the subject matter.
4) It permits readers to engage, give opinions, discuss, and offer their own expertise.
5) It will connect you to experts. You don't know everything. You will be shocked the number of experts you can integrate into your network.
6) Blogging on topic gives you a forum for direct feedback from the fans.
7) Recruits potential readers for your novel. As a general rule, if followers like and trust you for small,

incremental information, they are more likely to trust you for an entire novel.

8) Blogging adds content to the web, which builds your presence. NF (blogging on topic) does better with raising your search engine ranking
9) It will keep your brand focused.
10) It makes you be disciplined to write every week, thus keeping you and your subject matter top of mind.

**Remember, above all, content is for the reader** (reader is code for "customer"). This isn't about you. Remember, we as writers must **serve** the reader. If we don't, readers will gravitate to authors who do. This is true in fiction just as much as non-fiction. Fiction authors provide entertainment and escape. Readers like you for your content, and that isn't just a finished book (although ultimately it hopefully will be). If you write mysteries set in the 1800s and you blog regularly about that time period and mystery-information-factoids or whatever regularly, you will be in a greater position to already possess a following of mystery fans who respect your authority and talent to write on this subject. You will move from an unknown quantity to a known quantity much quicker than if you blog about writing or don't blog at all.

*****If you are unpublished and not yet agented, chapters/excerpts of your novel are NOT to a good choice for content.**

Unpublished authors who post sections of their novels (particularly works in progress) actually do more harm than good. In my opinion, it can lessen your chances of publication and can open you up to a world of hurt and possibly risk frustrating your followers. Permit me to expound.

**<u>The Problem with Posting Chapters On-Line</u>**

When you post sections of your book or novel on-line (in a blog, on your site, on your social media page) that content you post becomes public domain. Post too much of your content on-line and very few agents will want the full manuscript. For instance, I had to be very careful how I blogged about social media as I was writing this book. I wanted to generate interest and establish my expertise but I also ran the risk of giving so much away that people could just print off the blogs instead of buying the book.

Fiction can be the same. You could end up basically posting your novel on-line and for free. But that isn't even really the largest of your problems. By posting your fiction on the Web you open yourself up to all kinds of trouble.

There is a prevailing opinion that it is okay to post sections of your book that you can get feedback on your writing and possibly gain a following for your novel. Maybe. You should know by now that I am all about helping writers get content up on the web and teaching you how to use that content to gather a following of readers. The largest component to building a Web presence is that you must post regular content that is *informative, entertaining,* and ideally, *engaging.* But one question I get more than almost any other is, "Is it a good idea to post my writing on-line?"

My answer is, "Depends on what you're posting. Most everything yes, in limited quantity. Chapters of a novel? No. No. Definitely...um, no."

Posting writing on-line is helpful for certain kinds of writing and it certainly worked for the creators of *"Stuff White People Like"* and *"Julia & Julia"* (blog-to-book deals), but what about novels? Well, fiction does tend to always be the sticky wicket where the rules don't apply the same way. Most of the successes gained from posting sections of work on-line are from non-fiction, how-to, observational humor, etc. To my knowledge, there has never been an author signed to a book deal via posting chapters of a novel on-line. There have been self-published novels and text message novels (Japan) that have gained a following and publishing success, but even Jane Friedman (editor at Writers Digest) couldn't cite anyone who'd gained and Internet following through posting chapters or sections of novels when I addressed this topic in a recent blog.

Non-fiction and humor lend themselves to making good blogs and building an Internet following. But, for novels, many of the benefits of posting pieces of your novel break down, and I'll explain why.

**Test marketing.** There is a belief that posting on-line is a great way to test-market.

Fair enough. But before you get too excited, there are certain inherent problems with doing any kind of accurate test marketing for fiction. First and foremost, are you certain that you are getting an accurate statistical sampling when you post chapters of your book on your blog? Most of us cannot accomplish this.

In my experience, the majority of new writers do not have a statistically large following on their blog or even on social media. Because chapters of a novel are a piece of a larger whole, they are extremely difficult to gain the following and fan base like *"Fail Nation—A Visual Romp Through the World of Epic Fails."* In fact, "*Stuff White People Like*" had a Facebook following in the *tens of thousands* so it was easy to glean that it was popular and well-received. But chapters from an unknown, unpublished author? Tougher to duplicate these kind of numbers.

Any posted comments about your chapters are a tough way to gain any genuine insight because of this huge problem of numbers (or lack thereof). The smaller the group sampled, the less accurate the Bell Curve. Ten or even twenty people who take time to comment, positively or negatively is in no way an accurate litmus test as to how well your story is being received.

Additionally, the individuals who are most likely to follow or comment on the writer's work are generally a member of that writer's peer group—friends, family, fellow writers. Thus, it seems to me that this is the digital equivalent of telling an agent, "All my friends and family just love my book!"

Can you test market fiction by posting on-line? Sure. Anything is possible. But I think it is a lot tougher to do than it seems, and requires a very large and diverse following to get an accurate idea of how good your novel really is. Not to mention that a writer's work could look perfect and lovely when viewed in small snippets, but the novel as a whole, could be a disaster. I think there are better uses of a new writer's time and better content to use for platform-building than sections of a novel.

**Getting feedback on your work.** Feedback makes us better writers. But again, I think this is one of those ideas that are way better in theory than in practice when it comes to posting on-line. Sort of like, in theory I want my husband to tell me if I am gaining weight, but in practice? The plain truth is that we all have feelings and we all care deeply about our writing.

My issue with posting on-line is that it is a tough way to get accurate feedback for a number of reasons. When you get critique in your writing group, you know whose opinion is valuable and whose isn't. When an agent critiques your work, you know that is a valid critique whether you agree with it or not. But when you open yourself up to the worldwide web, who knows if that person commenting knows a protagonist from a potato?

Additionally (this ties in to my earlier point), if you have a network comprised of mainly friends, colleagues and family (which most people do), do you really believe they are going to be brutally honest and comment **publicly** that your writing was awful? They won't, because they aren't jerks. They are your friends and do not want to hurt your feelings.

It is one thing to ask for our brutal feedback in person, discussed over a table in a local library during critique group. It is a whole other ball of wax entirely when you want us to **post** that same feedback on the Internet *publicly and in writing*. Most of us just aren't going to do that to another writer, even when it comes to mild critique. If the writing isn't that great, most of us just won't say anything. And is that helpful to the writer for the purposes of feedback? Probably not.

But what about those who don't care about your feelings, who aren't personally vested in you?

Before you post anything, ask yourself one important question. **Can I take someone eviscerating my work in a very public forum?** Anonymity does weird things to people. Most of the time readers will be nice and kind and helpful, but sometimes they can be just plain horrible. If they tear apart a blog, that is one thing. That's 500-1000 words. Shake it off and move on. But with your novel? All it takes are a couple of negative remarks to crater your self-confidence and send even the best of us scurrying back to our laptops to rewrite our entire plot (and there might not even be anything wrong).

I remember a couple years ago I posted a humorous piece for public critique on my MySpace blog. I must have had 20 people who told me is was awesome and hysterical. But I had *one huge jerk* who posted a really hurtful mean comment, and I am still not over it to this day. I never felt the same joy about that article, and all it took was one person's nastiness to crush it. Was my response logical? No. But it was common. Humans are emotional creatures, and when you look up "Emotional Creature" in the encyclopedia, I think it says, "*See Writers.*"

Even published authors have a tough time when someone posts a nasty comment about their work in a public forum. But there is a difference. They have a published book, professional validation, and sales figures to ease their pain.

Another big problem with posting chapters on-line is that you could just irritate whoever stops by and takes the time to read your blog.

I know it sounds mean, but let me explain.

If you post sections of your novel, what you have posted are random scenes with random characters who your followers don't know from a hole in the ground. Not everyone followed your blog from Day 1 and even if they did, you are only posting snippets lest you give away the entire novel. Yet, you are expecting them to care. Our friends will care, but the network I am helping you build could number in the thousands, and you can't be buddies with all of them. So what about all those other people you desire to connect with? Will *they* care? I'll be blunt. Most of the time, they won't. They will move on to someone who blogs on topics contained to 1000 words or less, and save the chapters they read for *their* friends who are writing novels.

But, say you are an exception.

You have a really great story and writing voice and I, your follower, do happen to care about your new romance novel *Saharan Seduction.* I love Angelique and her fight to save the small African village from the warlord with the help of sexy Dr. Drake Carrington. I'm hooked. Ohhhh. But then I see that *Saharan Seduction* isn't finished? WHAT? Or, *Saharan Seduction* is finished, but you don't yet have an agent. That means I can't go **buy** *Saharan Seduction* so I can find out if Dr. Drake Carrington is able to successfully rescue Angelique from the rebel warlord. And not only can I not **buy** *Saharan Seduction*, but I also have no promise I will **ever** be able to buy *Saharan Seduction* since you don't have an agent, let alone a book deal in the foreseeable future.

Now I am no longer intrigued, I'm frustrated.

Excerpts from your novels are best used when you can deliver a product. Thus, if you have a book deal and can announce a release date, then feel free to post excerpts from *Saharan Seduction.* If you are already an established romance author and *Saharan Seduction* is your next work and likely to be published because you have an agent and a multi-book contract, feel free to post excerpts. Otherwise? Blog about topic. Blog about romance, true love, fate, rebels in Africa, or any other topic tangentially related that could get a readership interested in you and your work *Saharan Seduction*. Post love stories featuring Doctors Without Borders. Blog about similar books. Get creative. *Leave excerpts from your novel for later when you have a contract and a release date to entice followers to* ***buy*** *your book.* Save the bait for when it will translate into buyers.

Short stories and flash fiction do count as content, but I would limit how much you post. Why? Because it is tougher to get dialogue going with a piece of fiction on its own. If you post short stories or flash fiction, I strongly recommend that you post a question or questions at the end for the purposes of sparking discussion. The more engaged you get your readers the better.

**A Recap...**

By this point, you should have:

1) A brand name (name that will appear on your book)
2) A list of goals
3) An attractive headshot
4) A list of tags
5) A collection of content material—articles, reviews, photos, videos that all serve to establish your brand
6) A starting inventory of 10-15 articles (blogs) related to your topic and to serve as **strategic content**

What's next?

**Your Bio**

Your next step is to write three different bios using your favorite tags that describe *your brand.* I recommend one bio that is super-duper short (100 words), one that is super short (300 words) and one that is short (500 words). Yes, they must all be short. This is the 21st century. We don't have all day to read about your life, nor do we care. Sorry. This is marketing and sales. It is brief, catchy and to the point.

Most followers don't really care about your background save for the factors that will directly affect them. Nothing personal. We are busy and therefore self-centered. Just a reality. And in the end, it really isn't a bad thing. Don't you want people reading your work? Look at the bright side. The less time they spend reading about you, the more time they have to read your content. Your bio doesn't make money. Content is what will pay the bills.

In my professional experience, long bios actually do more harm than good. If you are lucky enough to get someone to care enough to even read your bio, you can only screw up. Why up your chances by making it long? Keep all bios short. Bios are not the place to establish your expertise or entertainment value. Your content does that.

I will use a non-writer example to demonstrate.

Recently I taught at a conference and there were at least ten agents there to take pitches from the attending writers. One bio in particular had to have been at least 1500 words, single-spaced, with lots of modifiers and no paragraph breaks. Guess which agent bio was skipped or left until dead last? (I actually polled my class) And the only reason anyone probably even read this agent's tome was because they paid good money to go to a conference and land an agent. If some writer read that agent's bio, I guarantee it was purely out of self-interest.

This agent's long bio clearly took her time to craft with loving care, but did it communicate the correct message? Instead of her extensive and elaborate bio painting the picture of a seasoned professional, this agent unintentionally portrayed someone who was insecure and not yet confident in her skill level. Why? There was too much extraneous information, a clear sign of overcompensation (padding the résumé). Did any writer really care where she went to college? Or where she had traveled in the world? Did we really want to know what her prior career choices happened to be? No. Those were all fluff and filler, and subconsciously we all knew she was green as grass. We wanted to know what agencies she has worked for and if she has been successful at closing book deals. That's it. If she didn't have a long résumé of success that was okay. But this agent would have been wiser had she spent those lines that told us about her world travel convincing us she was hungrier than the other agents, thereby redirecting her *newness* to her favor. New authors will need to do the same.

***After brand, the bio tends to be the next largest stumbling block for many writers, especially new writers.*** There is a good reason for this. Often, new writers don't have the confidence of a published author. Being published is a tremendous validation and boost to the ego. One suddenly feels confident enough to use declarative sentences and active verbs and write about one's subject of expertise. Suddenly you have a résumé that is full of your work and you aren't trying to pull gold out of thin air. But, it is going to take some time to earn your stripes. My opinion? Fake it until you make it. And if you must fake it? Fake it well.

## Bios for Non-Fiction

You should, to some extent be an expert in the area you have chosen to write about. If you are writing a book about financial planning, I would hope you have experience with a checkbook and a savings account or two. If you write about nutrition, then there is a background that led you down this path. Maybe you happened to once be morbidly obese and now you are a personal trainer. Or you just happened to be a mother of small children who one day noticed you needed a degree in organic chemistry to read the labels on food, and that bothered you deeply and profoundly and drove you to be a champion of healthy eating. It doesn't matter. You must have a special background that uniquely qualifies YOU to write on this subject. Merely having an opinion doesn't count. All of us have opinions, but your job is to convince others why they should care about yours.

It isn't necessary to have a PhD, but you must have some reason why you are different. Sell it in the bio. If you don't have a PhD or own a string of gyms, then you are going to need to tap more into emotional reasons why you are qualified to take up this mantel. Later, once you have written a string of blogs and articles and done enough talks at your local Rotary clubs, you will be able to modify the bio and gear it more to what you have *done.* A mother of an autistic child is just as qualified to write about autism as a childless pediatrician who specializes in neurological disorders. The mother of the autistic child will have an *emotional bio,* whereas the pediatrician will have a *professional bio.* Both have legitimacy. With the proper steps, the mother of the autistic child can add a *professional bio* to her *emotional bio* through her platform and activism (articles, appearances, speeches, blogs, etc.).

Whether you write a professional bio or an emotional bio, make sure you include the top tags you selected earlier in the body.

Bios for Fiction

Fiction authors always crack me up when it comes to writing their bios. It is so funny how fiction authors, who basically lie for 100,000 words, suddenly feel the need to confess the truth, the whole truth, and nothing but the truth when it comes to the bio and even the tags. Also, there is a lot of emotional distancing. Fiction authors have a tougher time believing that they are experts, especially if they haven't published anything or yet won any prestigious contests.

To illustrate what I am talking about, I asked an attendee of my last workshop if I could use her bios as examples. Karla is one of those bright, enthusiastic individuals whom you meet and *just know* she will go far with the right tools. After my class about tagging and branding, etc, Karla sent me her bio and tags, and she also chose to use a pen name **Piper Bayard** for privacy sake.

***Karla's first attempt:***

***Tags***

***Piper Bayard, seeds, apocalypse, post-apocalypse, bow hunting, Hospice, law, mystery, covert operations, assassin***

Notice Karla did mention her pen name, which was good, but then she NEVER mentions the words **writer, fiction, author** or anything to establish her as a writing authority or define her as an author. Karla gives tag words **hospice** and **law**. Why? You will see it in the bio. Those words describe **Karla (Piper) the person**. Karla feels secure as a person, but is shaky when it comes to envisioning herself as an author. Very, very common mistake among unpublished writers.

Let's take a peek at her bio.

***Piper Bayard***

***Biography***

***After growing up in a small town in New Mexico, Piper Bayard sat down at the age of 17 to "write what she knows," only to find she didn't know much. So she wrangled horses at a dude ranch, learned to belly dance, and achieved a degree in writing from the University of New Mexico while earning the label, "cancer survivor." Piper married a man who swept her off her feet, only to find her feet had been perched on the crumbling edge of a cliff. Result: personal apocalypse.***

***Piper's journey through her post-apocalyptic hell led her to aikido, the belly dancing stage, and law school at the University of Colorado where she began writing again. She was founding editor of the Student Trial Lawyers Association newsletter, The Gavel, and she co-authored "A People without Law," published in the Indigenous Law Journal. As with all epic journeys, personal and legendary, hers was made on blind faith, one step at a time, sometimes discovering green pastures, and sometimes tripping onto landmines.***

***Piper now lives in Colorado where she enjoys the fruits of her new world: a patient and honest husband, two smart, energetic children, and the opportunity to write fiction. Her first novel, SEEDS, is the story of a young woman who is battling a brutal, post-apocalyptic theocracy to win freedom for her people. It is the first book in the HARVEST TRILOGY. When Piper isn't writing, she is dancing, target shooting, SCUBA diving, visiting Hospice patients, or, of course, cooking for her family and chauffeuring her children.***

***So what does Piper Bayard know to write about now? Embrace the Apocalypse!***

Karla certainly gave a good effort, but there were a lot of problems. Hey, give her a break. It was her first try.

The single largest problem, first off, is that her bio is way too long for barely mentioning the fact that she is a writer or enough about the apocalypse. Karla gives us all of her growing up years and tells us about surviving cancer and all about her work in law and hospice, but most of that information does nothing to buttress her brand. It lacks any kind of focus.

Karla writes science fiction set in post-apocalyptic settings. Her personal apocalypse of a failed marriage sounds more like *Steel Magnolias* than *Blade Runner*. Karla was not writing a memoir or women's fiction, so that touchy-feely information actually did little to help define her as an author. What DID stand out, however, was Karla's interest in bow hunting. She also knows Aikido and likes guns. Those are skills and likes that are unusual for a female and for a typical writer. Those talents make Karla uniquely qualified to write this type of fiction. Such qualities are unusual and therefore make her stand out. They also serve to propel her image as an author of post-apocalyptic adventures with Tomb Raider-style heroines.

Karla has a common dilemma shared by many new authors. She isn't yet published. She doesn't have this long list of novels that help validate her as an author. Thus, she needs to try harder and be more creative in order to stand apart from all the other unpublished and even published authors. Karla actually was super creative and took a big risk with her bio, but I think she has created an image that is eye-catching, fun and daring. Karla has a wicked and quirky sense of humor that she also uses in her novels, so the bio fits very nicely with her writing voice.

***Karla's New Bio***

***Piper Bayard was once a happy and normal aspiring writer. She spent her days cleaning guns and practicing Aikido in between shuttling her children to and from school and crafting stories of sci-fi adventure. But she felt there was more. Then out of nowhere, White Sands beckoned with shocking visions of mushroom clouds and annihilation.***

***Something called to Piper day and night, whispering, "The end is near."***

***Piper left suburban safety and trekked through the New Mexico wilderness in search of that apocalyptic explosion. But, alas, she stood too close. Armed only with a ball cap, sunscreen, and her Maui Jim shades, Piper wandered through the desert for 40 days and 40 nights, wondering, "What the hell am I doing?"***

***By day, Piper followed mirages, leaving sand angels instead of footprints. By night, her nightmares returned. Mad Max and the Four Horseman of the Apocalypse until one night, the voices finally gave her the answer she sought . . . the key to the identity foreshadowed in her visions of the annihilating blast. ". . . and there came a Pale Writer. . . ."***

***The next day townsfolk found her, tube of sunscreen in one hand and a dried rib of saguaro cactus in the other, Piper scribbling madly in the sand. "Who are you?" they asked in wonder. "And why aren't you sunburned?"***

***To which Piper replied, "I am the Pale Writer of the Apocalypse."***

Karla is being very creative and very daring with this bio, but I think it is excellent. Will a bio like this work for everyone? No. But for Karla's subject, genre and voice, it is perfect. First of all, Karla (aka Piper) mentions straight out of the gate that she is a writer. No more emotional distancing. And notice how many great tags Karla now has with this second bio? Just glancing, I see **Piper Bayard** (her pen name), **sci-fi, adventure (**her genre), **writer** (her craft), **adventure/apocalypse** (her niche), **annihilating, Four Horsemen, explosion, Mad Max, blast, explosion, mushroom clouds (**all of these are prototypical apocalyptic imagery), and **desert** (standard post-apocalypse setting). This bio is chock full of tag words that will launch Karla straight to the top of her search when someone goes looking for fiction set in apocalyptic settings. People browsing the Internet generally aren't going to run a search on an author for what award they have won or what degree they have. They will run searches based on **content** and a bio like this will put Karla at the top of the list.

Also, the *Pale Writer of the Apocalypse* is a different and creative theme that tells anyone who visits Karla's site what she is about. It also opens up all kinds of possibilities for blogs, web imagery, promotions, etc.

Most anyone reading Piper Bayard's bio probably realizes that she didn't just one day leave her family to go wander in the desert on a vision quest. But isn't this much more interesting than the first bio? If you are a fiction author, ***you are a story-teller.*** Readers don't care where you went to college as much as they care that you can be entertaining.

You don't have to be as inventive as Karla, but remember you do need to stand out and your bio should be evocative of your writing voice. If you are artsy and whimsical, your bio should, ideally, reflect that side of you. If you are no-nonsense, cut to the chase, your bio can be as well. For fiction, I recommend humor, creativity or just some personal touch to make your bio interesting. Leave the résumés for the NF authors who have to demonstrate credibility. Résumés are not interesting. No one wants to write them and guess what? No one really wants to read them.

Your bio should have a little about you as a person, but anyone reading your bio wants to know about you as an author. I recommend only adding in personal information that supports the image you are trying to create. With Karla? Guns and Aikido were personal tidbits that served her brand, but law and hospice and thyroid cancer did not.

*Remember: Your bio should use the tags you compiled earlier and be clear, focused and catchy.*

Here is a real world example from one of my fellow writers on Twitter.

***Mary Rajotte: Mary's eerie tales inspire the urge to double-check the darkest closets, to anxiously peer under the creakiest of beds, and to always sleep with the lights on.***

I don't know about you, but this bio inspires me to want to read Mary's scary stories more than mentioning an MFA or a list of awards. It is creative and intriguing. Way better bait for a story-teller. Once you are published and can show off with all your titles, then rewrite the bio to reflect these achievements. I still recommend you employ the same kind of creativity, though. Résumés should be left to the NF folk in my opinion. If you write fiction your résumé is different. Your qualifications are that you can scare me, transfix me, perplex me, inspire me, whatever. So sell it in the bio.

Now you have a great bio (3 great bios to be exact). What is the next step?

## Time to Play Profiler

This is actually a step that seems natural for the non-fiction writer, but that is too often overlooked by the fiction authors. Non-fiction authors have to figure out their target audience and ways to reach that audience very early in the game (book proposal). I actually believe that it would be a great exercise for fiction authors to do the same.

Why am I suggesting you do this? Because not all social media sites cater to the same demographics. If you desire to make the most of your time and maximize your effect on social media, then you have to figure out who is likely to read your book. In plain English, "Fish where the fish are."

If you sell romance, what is your demographic? 31-49 year old women in a relationship (per RWA's site). What do 31-49 year old women do with their day when they aren't reading your romance novel? They go on diets and love Oprah, and Ellen and soap operas and knitting and scrap booking and gardening and parenting and American Idol and Dancing with the Stars. They do yoga and deal with stress and take care of elderly parents. This "other life" is part of what makes them long to read. They desire to escape, to be reminded that heroes triumph and love conquers all. In order to congregate with them, you first must figure out where they gather.

If you write fantasy and sci-fi, what do your fans do? Well, being a professional nerd for many years, I feel I can speak for my people. We go to Trekkie conventions and argue about whether or not it is possible to go faster than the speed of light (actual argument I had on the way to a Trekkie convention—swear to God). We play X Box, World of Warcraft, PS3 and Dungeons and Dragons. We love comics, cult movies, and X-Men and Star Wars and quote Monty Python far more than is socially acceptable.

If you desire to connect with your readers, you must go to their favorite watering holes. When we start thinking like a fan and not a writer, THEN we will find our readers. That is why profiling is critical.

And beyond the reader, profile the purchaser. If you write books for children, who will buy the books, and where do these people hang out? Who are the major influencers of your consumers' demographic? In this case, it might be teachers, librarians, or home school networks.

Take a few minutes to jot down your demographic and then envision their day. Write down preferences, activities, hobbies, and anything that seems salient. This might feel a bit silly right now, but it will be vital for later when you get started on social media. My goal is to make you better than your competition. Think like a reader instead of a writer and your time will be far more productive.

**Recap...**

By this point, you should have:

1) A brand name (name that will appear on your book)
2) A list of goals
3) An attractive headshot
4) A list of tags

5) A collection of content material—articles, reviews, photos, videos that all serve to establish your brand
6) A starting inventory of 10-15 articles (blogs) related to your topic and to serve as **strategic content**
7) Three different bios. 100 words. 300 words. 500 words.
8) A detailed profile of your reader & consumer demographic

Now we are ready to ease into the technology. For the purposes of not making anyone's brain explode, I am only going to address the top three platforms—MySpace, Twitter, and Facebook in addition to Wordpress blog site. This walk-through is going to be very basic so that you have a firm grasp of fundamentals before you try to do anything fancy (that is for another book).

So let's get started.

**Getting Familiar with Social Media Terms**

Social media is a world of its own with its own vocabulary. This section will give you an overview and understanding of the most commonly used terms.

**Social Media Optimization (SMO)**—Basically, SMO makes your content linkable, easier to tag and bookmark, helps to reward inbound links, increases the reach of your content, and encourages the mash-up (a Web page or application that presents a hodgepodge selection of the best of the best, which will increase your influence).

SMO helps your content travel and makes it far more user-friendly.

**Tag**—metadata (data about data) that helps locate content on the Internet.

**Bookmarking**—a method by which users can share, organize, search and manage content.

**Stream**—a streaming update on your MySpace that lets you know what others in your network are doing—new pictures posted, new status update, new video, blog, etc.

**Newsfeed**—Facebook's version of the stream. Updates on what everyone in your network is doing.

**URL**—stands for *Uniform Resource Locator*. You will likely never need to know what URL stands for, but you should know what it is. Basically, the URL is an address so you/your content can be found. Like a house has an address for its street, sites have addresses on the Internet. www.kristenlamb.org is my home address when it comes to my site and the Internet.

**Blog**—is a term that came into existence in the late 90s. It is a combination of 2 words, *web* and *log* to form *blog*. All blogs are websites, but not all websites are blogs. What makes the difference is in the presentation. Blogs—which may consist of diary entries, text, photos, audio/video—are generally updated frequently and, if done correctly, regularly. By contrast, a static website normally serves as a resource for information that doesn't continually change.

**Hyperlinks**—are digital breadcrumbs. This is when the address of a site is embedded in text posted on the Web. For instance, when I blog on certain topics, I will refer to other sites that had good information that buttresses, explains, or expounds on my point. I will also give lists of recommended sites. When people click these sites (often colored), it takes them from my page, to the web page I was referring to with one click.

**Pingbacks/Trackbacks**—These notify the author of the site that someone has included a hyperlink to his site. For instance, my Wordpress dashboard will notify me when anyone references my Warrior Writers blog in one of their posts. This does three things. First, it extends my digital reach. Trackbacks/pingbacks are a way of linking blogs together and widening the web. Secondly, trackbacks provide me an opportunity to thank the person who took the time to reference my blog and direct her traffic to my site. Thirdly, I can make sure my site is being referenced in an appropriate way, not being misquoted or misconstrued. It provides me an opportunity for rebuttal if need be.

**FEAR**

***I want to share an interesting industry blog I ran across when I got a recommendation from a friend. The Warriorwriter's Blog has a wonderful post today about fear. I think it's something every writer has dealt with at some point in their career. It really hit home with me this morning because I've been struggling...and not sure why. I thought maybe it was because my life is so chaotic with different directions I'm being pulled in right now. That may play a big part of my issues with not writing, but I think fear could be part of it.***

I will teach you how to use these in your blogs to maximize your reach and to edify others and their hard work.

**Status Update**—MySpace, Facebook and Twitter all have a format for **microblogging**, known as the *status update*. Basically, this is a small window where the user (you) answers a very simple question, "What are you doing?" (Twitter) What's on your mind? (FB), or "What do you want to share?" (MySpace). The status update is invaluable for connecting with others, creating a network, and also getting one's content out on the web (we'll talk about how later).

**Tweet**—basically is a synonym for *status update.* Twitter, unlike the other two major social networking platforms, focused on one aspect of the social networking experience—the status update—and built a social media *tour de force.*

**Retweet/Repost**—are essentially synonyms. Status updates are one of the most vital components of social media. Remember earlier we talked about the reciprocity factor? When you post information, your reach is only so long. It will be in the connections you make and the inherent value they place on your content that will make the critical difference. When people *retweet* or *repost* your information, your digital reach extends exponentially.

***TWITTER RETWEET IMAGE***

******Just these two super nice ladies RTing my blog extended my reach by over 2000 people.***

**SIM**—stands for *Social Influence Marketing.* Unlike traditional marketing and PR, SIM seeks to influence the consumer *and* his/her community/peers. SIM is much more social and community oriented.

**RSS**—stands for "Really Simple Syndication." This feature allows others to essentially "subscribe" to your blog, so they don't have to keep going to your site to see if there is anything new. RSS feeds also replace e-mail newsletters. So, for instance, I always post once a week on my blog, but which day that week is often anyone's guess (yeah, *Note to Self—get better about that*). Anyway, RSS makes it simple for anyone to follow the blog.

*RSS FEED IMAGE*

Simply hit the orange button and it will show up on a Web newsreader.

**Newsreader**—is basically the place where all the cool stuff you find on the Web can be funneled so you have it all in one handy place. I use Yahoo mail, so what happens is that every blog I subscribe to via RSS now shows up on my Yahoo home page.

And no need to panic if you aren't tech savvy. The first time you hit that little orange button, a menu will pop up asking you where you would like the RSS feed to go. I suggest a Web feed like Yahoo. Otherwise, likely a list of other options like FeedDemon will be offered and you just download and follow the cues along like most other software applications.

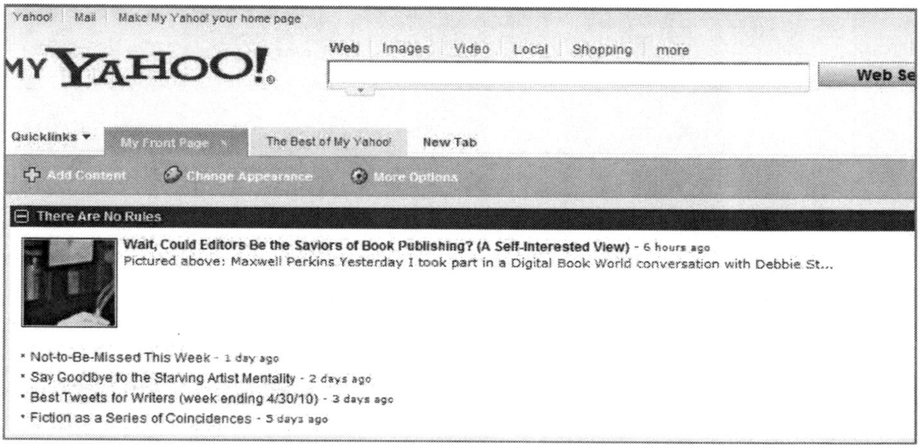

***YAHOO NEWSREADER IMAGE***

**Domain Name**—basically your "address" on the web. Likely it will be your brand name (the name that will appear on your books). For instance, www.bobmayer.org.

**Widgets**—basically widgets are portable chunks of code that can be embedded in a larger HTML page. Think of widgets this way. They are like refrigerator magnets that serve a specific function stuck to your fridge (that has its very own function). One magnet could be a calendar, a bottle opener magnet, a chip clip magnet, or an easy reference to Poison Control, and they all hang out on the front of your fridge for the purpose of making some task easier, yet their existence doesn't change the purpose of the refrigerator.

Widgets will make it easy for people to interact with your content. An RSS widget, for example, will help people subscribe to your blog. It is an additional button with the additional purpose of making some task (Social Media Optimization) easier.

**CAPTCHAs**—are used to decipher between humans and computers and keeps spammers off your page. **DON'T USE THEM.** They are seriously annoying to anyone who is trying to post on your blog or befriend you on MySpace. Yes, they keep out SPAM, but you should be tending your pages regularly and you can just manually delete any SPAM. With the current technology, most of the time, SPAM will be put in a separate box for you to scan and delete. Make it easy for people to interact with you, so forget the CAPTCHAs.

**Content**—makes up what you add to the social media experience. Content can be images, videos, pictures, comments, blogs, links, whatever. The more content you provide the better, namely because it helps foster discussion. If no one posts content, you become like a room of twenty people standing around and no one has a thing to say. Content sparks dialogue. Dialogue eventually translates into relationships, and relationships into community, and community becomes platform.

**Critical Mass**—the number of individuals in your network to make interaction meaningful. For instance, Twitter is a great place to get answers to questions. So if you post, "Which site is best for web hosting?" and you have three people in your network, you are severely limited for obvious reasons. But, if you have 200? The odds of getting a good answer tremendously increase. There is a certain number one must reach in order to maximize the social media experience. That number fluctuates depending on the goals you have set.

**Application Software**—or application or app. Fancy word for a *tool.* Application software is a subclass of computer software that utilizes your computer's abilities for a specific task. Think about it this way. Gasoline really isn't much use until it is *employed by something* for a certain task, like a car engine (used for the purpose of locomotion). The main application we will discuss is called TweetDeck and it is used to organize information effectively. If you follow 5,000 people at a time, you have to have a way of making it manageable.

**Groups**—On MySpace and Facebook there are groups of people who converge to discuss topics they share in common. Think of these groups like clubs on social media. There are all kinds of groups. Political groups, activist groups, history lover groups, etc. The sky is the limit. If you can't find a group of people who want to discuss your topic, then it is easy to start one. For instance, "Physics Doesn't Exist, It's All Gnomes" or "Writing Papers Single-Spaced First Makes My Double-Spaced Result Climactic" are actual Facebook groups. Just these two groups collectively have almost 200,000 followers. But the appeal is clear. It is a place for people to gather and be silly, creatively silly. Yes, I am a member of both groups.

MySpace is very similar. Most of your networking should be done in these groups in that these are the places your readership is congregating in their free time.

**Hashtags (#s)**—are a method of starting conversations on Twitter that are not limited to one's personal social graph. Think of the # as Twitter's answer to groups (discussed above). A hashtag is similar to a Group on MySpace or Facebook in that generally people congregate based on a shared belief, idea, subject or goal.

On Twitter, most of the time we engage with people we are following or who follow us. We also engage with friends of friends. But what about all the rest of the Tweeters out there? We like to talk with others interested in the same topic, and hashtags help make that happen. #writegoal is a popular hashtag conversation for writers. Place this hashtag at the end of your post and it will appear in a column that everyone else in the hashtag conversation can see. #writegoal was created by talented romance author and wonderful human being Anna DeSefano so writers could have accountability. Writers from all over the world post their daily goals and then find answerability with their fellow #writegoal participants.

So if I wanted to participate, I would tweet this:

**Goal is 1000 words today #writegoal**

That *#writegoal* directs my tweet to the #writegoal conversation. Now I can interact with fellow writers from all over the world, whether they happen to be in my network or not. They don't even have to know me to bust my chops when later I post:

**Got distracted watching History Channel and didn't write #writegoal**

Writers from all over the globe can now gather together to remind me that watching TV will not get my books written. Conversely, when I reach or exceed my goal, these same writers can also offer congratulations and encouragement. No one is going to drag me to writer jail for not making my goal, but it is funny how just this small measure of accountability can really kick one's derriere into action.

***#WRITEGOAL HASHTAG CONVERSATION IMAGE***

Hashtag conversations can multiply your influence exponentially. Other popular writer hashtag conversations are #pubtip, #amwriting, #nanowrimo (Stands for National Novel Writing Month).

Hashtag conversations can be permanent or only take place for a short period of time. Recently, a group of NY literary agents started the #askagent conversation. During a certain window of time, writers from all over could ask questions of the participating agents simply by using that hashtag.

***#ASKAGENT HASHTAG CONVERSATION IMAGE***

We will discuss hashtags a bit more in the Twitter section. Hashtags are going to be invaluable for finding future readers.

**TweetDeck**—a popular, downloadable application designed to make it easy to follow numerous categories simultaneously. Like hashtags, we will discuss this in detail in the Twitter section. So breathe.

**Users**—anyone participating in social media. Ah, but not all users are created equally. They are divided into categories that correspond with the influence they exact of their surrounding networks.

**Expert Influencer**—is just what it says. These are the authorities in a certain subject, and people look to the experts for information, advice, and guidance. The experts are heavyweights when it comes to influencing the decisions of those in their networks. *Expert influencers* usually have a picture of themselves as their icon. They also generally have huge following that number in the thousands or tens of thousands, depending on the platform. Also, a quick glance to their website (which is usually denoted in the bio) will give you a clear picture that this person is an expert in her field. Oprah. Enough said.

**Referent Influencer**—is in the person's social network and exercises influence. *Referent influencers* are a little trickier to figure out. They generally have a fairly large following, but not always. Which person has more impact? User A's network is comprised mainly of her friends and family and people she knows from back in high school. User A has a lot of followers, but most of them are not active on social media and half barely know how to log in to Facebook. User B's network is relatively small and comprised of known experts (Oprah) or other referent influencers. Quality and quantity not the same thing.

So how do you figure out the referent influencers? Well, you have to participate so you can pay attention. For the most part the **referent influencers are highly active on social media** and thus usually have a larger following than the casual user, but maybe not as large as the expert. Yet, it is their level of *meaningful activity* that makes them essential to have in your network. They post a lot of times a day and are well-known, liked, and respected for good content. People around them trust them for good stuff. These are the people you miss when they take a day off.

In my opinion, the *referent influencer* is the most valuable. Why? First, **it is easier to get close to them and befriend them and gain their support.** If you write a blog about parenting, what are the odds of becoming part of Oprah's inner circle? Referent influencers are far more approachable.

Secondly, **referent influencers are genuine and personal and thus exercise tremendous authority**. I think that people tend to trust these types influencers almost as much the experts, if not more. Why? Well, human nature. We like things from the proverbial mouth of the horse. We can't really be sure Oprah, and not some staff writer, does her posts. But, Jenni Carpool, mother of four and hockey mom who also blogs about parenting, always posts good stuff and has 3000 people in her network is easier to win to your side. It is a much easier feat to get Jenni Carpool *to repost your blog* about dealing with teenagers than Oprah. Additionally, those who follow Jenni

Carpool know she is an authority and listen to her much like an expert, even though, by strict definition, she isn't.

Thirdly, **there are more referent influencers than expert influencers. A lot more**. There are far more Jenni Carpools to befriend than Oprahs.

**** Many referent influencers are considered experts in certain subject areas. Pay attention.*

**Positional Influencer—**is often in the person's inner circle. Friends, family, spouses are all examples of positional influencers. Yes, whether most of us admit it or not, our mothers' opinions still influence us.

Virtually everyone on social media is a positional influencer to someone else. Positional influencers can be very valuable to a writer, especially in certain genres. For instance, I imagine that most 4-year-olds don't drive down to Barnes & Noble, slap down a credit card and buy a stack of kid's books. But moms do. If you happen to write for children, middle grade, teens, or any group that typically would not be the purchaser of the book, then you must target the positional influencers or risk losing a huge percentage of your potential consumers.

We already talked about profiling the reader as part of your social media campaign, but one would also be wise to profile the purchaser.

***Ideally, you will recruit the referent and expert influencers who hold sway over the positional influencers.*** Recruit Parents Magazine or Jenni Carpool to your side and the moms will listen.

**** The key to doing social media well, lays in recruiting and mobilizing the all types of influencers, particularly the referent and expert influencers.*

**<u>Ways to Influence</u>**

Earlier we discussed content, but how you plug your content into the whole of the Internet is what will make the crucial difference. There are a lot of ways to influence those around you, but let's look at the most popular for building an author's platform.

For writers there are four top ways to influence.

## Blog

Post regular content that is informative, entertaining and engaging. Yet, the rules differ for non-fiction and fiction, so let's address these two separately.

### Non-Fiction Blogging

We have already discussed this earlier, so I won't beat you over the head with this...too much. Blog on topic. This is easy for the non-fiction authors. Non-fiction authors are there to inform, educate, and serve the readership in some way. If you want to publish a diet book, blog about healthy eating and logically related topics like exercise and beauty. Duh, right?

Unlike a fiction author's blog, I don't believe it is as crucial to leave the end open for questions and comments. If you do, be careful how you approach doing this. Open-ended questions can establish dialogue with readers, but ask the wrong questions and you could shoot yourself in the foot. When it comes to NF blogging, **you are the expert**. Most people go to a NF book or blog to be informed or to solve a problem. Ask for feedback the wrong way and you unwittingly could undermine your position as the expert.

For instance, if Rachael writes a diet book based on the principle that low-carb is the way to go then asking for feedback at the end might be risky because it makes her seem unsure of her position. "Low carb is the best way to lose weight safely and effectively. What do you think?"

Yet, if Rachael just writes the blog in a declarative voice, then comments about her blog run less risk of arguing with her position because she has not asked for an opinion. NF isn't subjective like fiction is, and thus contrary opinions could confuse readers and challenge the author's authority on the subject. If you are selling a book on dieting, financial planning, or even social media, you are selling **a program based on quantifiable results you have achieved**. For you to open yourself up to other methods of approach on a site designed to help your platform is risky.

Fiction doesn't require you to be an expert on dragons to write fantasy, but NF does demand that you be an expert of financial management if you desire to write a book telling others how to manage their money. It is vital to always be seen as that expert. No one wants to buy a book from someone who appears wishy-washy or not to know what she is talking about.

This is one of the main reasons it is imperative to be active with your blog. Negative or contrary statements are fine and should be left in the comments section so long as they are appropriate and not abusive. But you will fare better if you rebut quickly and put the readers at ease that you know your subject and are okay with differing opinions.

So if you want to ask open-ended questions to get a discussion going, do so in a way that keeps you in the expert position. Why create unnecessary headaches? Much like a lawyer is told: Never ask a question you don't know the answer to during a trial. If you're the expert, you don't want to get into an argument in the comments section on your blog

What is a good way to drive traffic to your blog?

Remember that little gizmo called a **hyperlink**? This is the place that **hyperlinks** and **trackbacks** will help maximize your reach. Think of hyperlinks like...name dropping. (Trackbacks/Pingbacks notify the person their name has been dropped). But better than just name-dropping, these hyperlinks are like digital breadcrumbs that will thread people together and widen-widen-wiiiiiden your reach. You can do this as easily as, when you write your blog and reference this person, you type in their web/blog address in the body of your work. It should turn blue, cuing those who read that they can click that and go to the blog or article you referenced. You likely see them and use them all the time, just didn't know what they were called.

Maybe, though, for appearance sake, you don't want to type in a full web address. No problem. When I walk you through Wordpress later, I will teach a way of attaching the link to a word or name so as not to interrupt the flow of your writing, but still gain the benefit of referencing other blogs/articles/experts.

Hyperlinks can make a huge difference in how far your message travels. For instance, recently author Jody Hedlund read my blog about branding and LOVED it. She loved it so much that when she later wrote a blog on a similar topic, she inserted a hyperlink back to my page. So when her followers read her blog and they were excited because she was excited, they could click the hyperlink and go directly to the blogs she referenced.

In fact, Jodi was so clever (she is a very clever girl), that she inserted multiple hyperlinks. Every time she made a point, she inserted a hyperlink to an article or blog by experts who supported her opinion. This action helped support the opinion of her readers that Jody was knowledgeable because she demonstrated very transparently that she had done her homework. When we write NF, we frequently use citations to buttress our ideas, opinions and conclusions. Hyperlinks help reinforce your image as an expert and at the same time, improve your searchability and the searchability of others.

On the other end, people like me were notified of Jody inserting a hyperlink to our site via a trackback. So we were able to then go see what had been said about our blog, comment, and even drive our own readership to share in this accomplishment.

Aspiring writer Jan O'Hara actually didn't know me at all, but she happened to follow Jodi's blog. She clicked on the hyperlink and loved my blog just as much as Jodi promised. Jan then not only followed the advice in both blogs, but she inserted hyperlinks into **her blog** about changing her moniker to her pen name and why. Now, my blog had exposure not only to my network and to my readers, but now also was exposed to Jody's AND Jan's networks. Just by one writer being thoughtful enough to insert a hyperlink, my blog's exposure suddenly multiplied exponentially. Now when people visited my blog it was because someone else edified me. Can't beat that with a stick.

The more hyperlinks you employ, the better. It will not only increase your influence, but it will help those who helped you by sharing their knowledge. This is just another way that we writers can help each other expand our platforms. I like to describe it this way. Each writer's social media platform/blog is like an individual quilt square; intricate, beautiful and unique. Yet, alone this one square cannot fulfill its purpose. A lone quilt square is good for what? A coaster? A pot holder? Yet, stitch that quilt square to other equally special and unique squares and then you are on the right track. Stitch enough squares and the purpose can be fulfilled (bedspread/blanket). Stitch even more squares and the purpose can be exceeded (story-telling quilt, quilts to teach cancer awareness, art).

Our social media platform is only so good by itself. It is when we effectively "stitch" our platforms to others using tools like **hyperlinks** that we collectively become greater and greater and greater until we exceed anything we could have ever achieved on our own.

The **trackback/pingback** is what will alert you that someone has employed a hyperlink using your web site, article or blog. Now you can thank them, comment on their blog, or read what they had to say and what you wrote that was so interesting they felt the need to reference it.

I love getting trackbacks. They are good for the ego and the soul.

### Blogging for Fiction

We already discussed why chapters of a book are a bad choice for content. If you decide to post short stories or flash fiction, make sure you include, at the end, *something* that engages the reader other than, "Hey, what did you think?" My advice it that you treat this like a book club. Ask questions that make the readers of your blog have to think, form an opinion and desire to respond. Questions that encourage participation and a dialogue are invaluable. Fiction is subjective, so capitalize on this.

***Tip—One of the reasons I recommend blogging on topic at all times is that one has to take into consideration the way that search engines work in conjunction with user behavior. Author Frances Hunter posted this on my blog, "Is it a Good Idea to Post Chapters of Your Blog On-Line to Build a Platform?":

> ***It is also worth noting that fiction does not rank very well in search engines, because it does not contain the kind of keywords that cause it to be picked up. In other words, people who might love your work cannot find it to comment, give you feedback, or become your audience. The posts we have done on our blog that are book excerpts are among the lowest ranked posts we have ever done, because no one can find them.***

One needs to appreciate that, at least for now, most of us go to the Web for information (the domain of the NF writer), and it will take time for people to seek entertainment in the same fashion. This is why I strongly advise that the blog on your content is the bait and hook to draw in a reader, and the fiction is extra. For those who use a web site, just make an additional page for fiction samples, but use your blog to draw in the numbers.

### Tips for Blogging in General

Try to respond to any comments on your blog. If people are nice enough to take the time to comment, thank them individually, or, if you have too many (great problem) then just post a "thank you" to the comment-makers as a whole.

Blog regularly. Some experts advise blogging as often as possible, several times a week or even several times a day. Well, it is up to you, but I have to work hard enough to be entertaining, informative and engaging once a week, let alone once a day.

My professional opinion is this. I really don't care so long as you ***post regularly and at least once a week.*** Who is going to remember to follow a bi-weekly or a monthly blog? By the same token, who has the time to read the same blog three or four times a day? I prefer quality over quantity. It is ideal that you post on the same day each week, but sometimes that is just tough to do. Just make sure you post every week, whether that post is Monday or Friday shouldn't be as big of a concern as just ensuring you post content *some time* during that week.

Blogging is really author training. It makes you write good content on a regular basis. But, it can be like going to the gym. Pace yourself and set achievable goals. I would rather see you post well-written, engaging content once a week, than fill up the web with foolishness. Remember, people hone in really quickly who posts good content. A reputation is hard to earn and easy to wreck. Make sure you always focus on the big picture.

Fiction writers can also employ hyperlinks. Have a friend who wrote a great short story, blog or commentary? Reference it in your blog. It will permit your followers to find other material just as good as yours, and they will appreciate you for it. Other writers eventually will catch on and hopefully do the same for you. Talk with your friends who write the same genre and see if you can't use this book and work together to support one another.

Other than hyperlinks, how do you drive traffic to your wonderful blog? Here is where the *status update* becomes your new best friend.

## Status Updates

The *status update* is just one more reason to participate, participate, participate. I cannot emphasize this enough. How can people get to know you if you are never around? In social media, you will only get out of it what you put into it, and proper use of the status update is fundamental. The status update is probably the most powerful tool in your arsenal. It is what you will use to connect with, befriend and mobilize your influencers—expert, referent, and positional.

**The most effective use of the status update is as follows—equal parts of information, reciprocation, and conversation**. Earlier I showed you a "tweet." Status updates look the same. They are how you communicate to others and direct them to static web pages, other social platforms, or blogs.

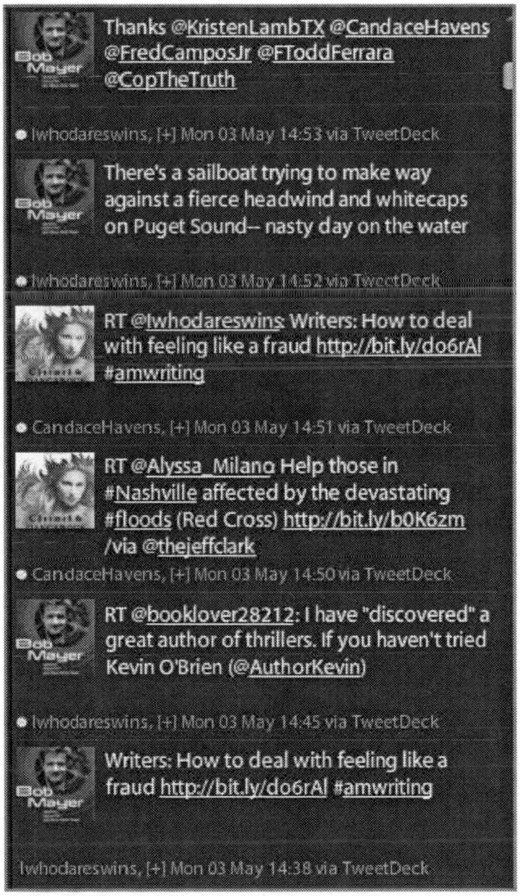

Thanks @KristenLambTX @CandaceHavens @FredCamposJr @FToddFerrara @CopTheTruth

lwhodareswins, [+] Mon 03 May 14:53 via TweetDeck

There's a sailboat trying to make way against a fierce headwind and whitecaps on Puget Sound-- nasty day on the water

lwhodareswins, [+] Mon 03 May 14:52 via TweetDeck

RT @lwhodareswins: Writers: How to deal with feeling like a fraud http://bit.ly/do6rAl #amwriting

CandaceHavens, [+] Mon 03 May 14:51 via TweetDeck

RT @Alyssa_Milano Help those in #Nashville affected by the devastating #floods (Red Cross) http://bit.ly/b0K6zm /via @thejeffclark

CandaceHavens, [+] Mon 03 May 14:50 via TweetDeck

RT @booklover28212: I have "discovered" a great author of thrillers. If you haven't tried Kevin O'Brien (@AuthorKevin)

lwhodareswins, [+] Mon 03 May 14:45 via TweetDeck

Writers: How to deal with feeling like a fraud http://bit.ly/do6rAl #amwriting

lwhodareswins, [+] Mon 03 May 14:38 via TweetDeck

***STATUS UPDATE IMAGE***

This is an excellent example of how social media ideally works. Time on Twitter flows as such—the oldest posts are located toward the bottom, and the most recent are positioned at the top.

@iwhodareswins is NY Times Best-Selling Author Bob Mayer. Yes, even Bob used to use a moniker @iwhodareswins. Bob has since listened to my teachings and changed his username to @Bob_Mayer proving once more that the truly successful never stop learning.

I digress...

But what I really want you to look at is this Twitter conversation. Bob starts with a tweet (status update) about one of his blogs. Bob sells books on the craft and also teaches writing workshops around the country. This particular blog targets his potential audience/consumer—writers. Next, Bob RTs someone else and, in the process, edifies another author. A totally different author from the one Bob promotes, well-known and much-loved Candace Havens, RTs Bob's information in kind. Then Bob tweets something personal about his life to connect to others as *people.* Finally, Bob takes time to thank everyone who RTed his information. Perfect form!

***Information*** *and* ***reciprocation*** go hand in hand. If all you post on Facebook or Twitter is "Look at me! Come read my blog! Come read my book! Look at my pictures! Read my poem!" you will quickly become about as effective as SPAM. People who use social media are there for a sense of community, and they will catch on very quickly if you are only there for your own self-promotion. Likely they will ignore you or just boot you from their network altogether.

***Conversation*** is just that. Have a dialogue with others. Establish rapport. If all you do is post and repost links, you become a "bot" and you lose tremendous influence.

**You should strive for a balance between information, reciprocation and conversation**. The figure above is an excellent illustration of the perfect balance.

## The Importance of Conversation

All three factors are equally important, even the chit-chat. There is a massive misconception that idle chatting has no place in platform-building, but that couldn't be farther from the truth. One thing I hear over and over from those who don't yet understand social media is what a waste of time it is to tweet or post mundane things like what we had for lunch. Not at all. It is a waste if that is ALL you tweet or post.

Yes! We do care what you had for lunch. Let me explain it this way.

Why do people mention the weather when trying to strike up a conversation? Um, because the weather is universal and quickly establishes a non-threatening common ground to build rapport. Same with posting about lunch or Starbucks or our pets.

Bob Mayer actually did something very brilliant when I first introduced him to Twitter. Bob would tweet lines from movies like *Casa Blanca* and *Ghostbusters*. It was amazing to watch my TweetDeck light up as all those following Bob wanted in on the movie quotes. It was silly and fun, but it made Bob human, approachable, and established common ground with his fans and other not-yet-fans who just happened to be passing by and wanted to quote movies too. And I just know every one of Bob's fans was just glowing because their hero was interacting with them.

People are on social media for one huge reason....connection. Humans are social creatures. Without community we suffer mentally and emotionally. Humans left alone often go crazy or die . . . or go crazy *then* die. I think it can also be linked to why introverts are just a more than a little odd. KIDDING!

Yes, we want content. But we also like chit-chat mixed in. My favorite people to follow are the ones who permit me a peek into their lives. Those glimpses are memorable.

Pictures of their family dogs. An update on a desk from IKEA that is half-way put together. An update of a desk from

IKEA that is all the way put together . . . and there are "extra" bolts and screws.

What I just mentioned was content from Bob Mayer's blog *two years ago*. I had known about Bob the Big Author for years, but knew next to nothing about Bob the Person until I ran across his web site and read a section of his blog called "Bob Land." The story about the IKEA desk made me laugh, and I remember it to this day. Why did Bob's posts strike a chord? Because I also have a dog, and I too know what it is like to put stuff together and have "extras." Those posts made Bob human. In fact, he will even tell you that to this day, his biggest hit blog was the one about his dog Cool Gus and the kayak. Bob got the idea to take Gus along kayaking. Gus spotted a whale and took off, dumping Bob in Puget Sound. The blog was a huge success!

Bob Mayer isn't the only one I have seen employ this tactic successfully. Teresa Medeiros posts pictures of her cat *Buffy the Mouse Slayer.* Susan Wiggs posted pictures of her daughter's wedding. Christina Dodd shares funny pictures from conferences and asks followers to help her figure out titles for her upcoming books. James Rollins tweets about his trips and about his new gadgets. Candace Havens posts clips of her son singing opera, and Rosemary Clements-Moore jokes about her battle to exercise and avoid sweets. All of these people are big authors. Some are even NY Times best-sellers. Do these posts tell us anything about their books? About writing or the world of publishing? No. They do far more.

These posts give followers/fans something in common with people, who, in any other place, would be very unapproachable. These personal slices of life make them less superhero, and more like us, and we like people who are like us. It makes an author "human" and gives something innocuous to gather people together in COMMUNITY.

I firmly believe this tactic is even more important as authors get bigger. We fans get this weird vision of your life, and it rarely looks like our lives. When it does? Wow! We LIKE hearing that your dog ate your glasses or that you are stuck in line at Costco. Why? It makes you real and we can relate. And, like the weather, it gives an easy way to strike up conversation. When we converse, even briefly, it creates relationship and a feeling of obligation. If faced with the choice of two equal books, we will choose the one written by the author we talked to on Twitter about how it is impossible to leave Wal-Mart for under a hundred bucks.

Conversation also does another thing. It gives others a break from us being constantly pitched to. If all we hear is, "Go to my blog! Buy my book! Sign up for my workshop! Go join my fan page!" we eventually tune it out because it becomes like all the other things and people pulling at us, wanting our energy and attention. It becomes SPAM or the commercials on TV. White noise. We disengage.

**Conversation is your most powerful tool of connection.**

Most of us live lives where it is easy to feel very unimportant and overlooked. We talk to computers when we bank, talk to machines when we shop, and no one seems to even remember our name. We love kindness in all its forms and we remember it. We like to be noticed, even if an author, or even a regular person, just takes two seconds to reply and say, "My dog once ate my couch."

Remember in the first part of the book we discussed the *likeability factor. Many of us will buy from who we* ***know and like first****.* It is this interpersonal connection that is so vital to driving sales. Yes, conversation can turn into relationships, which ultimately translate into books sold.

Recently on Twitter, out of nowhere this romance author named Kay Thomas complimented my Warrior Writers blog and posted the link to my blog. The only reason I saw it was it popped up in a column I have created for any time my name is mentioned (will teach this later). At the time, I wasn't even following Kay, but such an unexpected act of kindness propelled me to action. I added Kay to my network, and we chatted back and forth about my blog. I scanned her bio and saw the name of her latest romance novel, *Bulletproof Bodyguard.*

The next day, and I am not lying, I saw Kay Thomas's *Bulletproof Bodyguard* at Borders. I was wandering the bookstore looking for the computer section when I spotted the title and remembered it from Twitter. Kay had been so nice to chat and post my blog that I felt compelled to return the favor and *buy her book.* Out of all the books on all the shelves, I gravitated to *who I knew and who I liked.* Surely I am not the only one on the planet who would do something like this.

## Three Rules of Status Updates

### Rule #1—Make sure that content you post is entertaining or informative (Information)

Other users in your network will rapidly hone in on who posts good content, versus those who just toss out fluff and junk. The Internet is an overwhelming place, with more information than we could ever consume in a thousand lifetimes. We also are rightfully scared of surfing alone. It's too easy to stumble into sites that can make us the victim of spyware, viruses or phishing. Thus, users actively seek out gatekeepers who they can count on for links to good, safe content. Make sure that is you.

### Rule #2—Make sure you do unto others as you would have them do unto you (Reciprocation)

Social Media is a world of quid pro quo. Those who forget this are swiftly penalized. It really boils down to common sense and good manners. You want people to go to your blog that you sweated over and crafted with blood, sweat, tears, and dreams, right? Why would anyone else be any different? Make sure you repost the works of others, and make sure you still follow rule #1 and take a few seconds to at least scan the blog to make sure it is something you want your name attached to.

I am on Twitter quite a lot and I read virtually every blog I repost. There are certain individuals who I feel comfortable reposting immediately because they have established a reputation for quality information. Now, mind you, I do generally make sure to read the blog at some point during the day because that is one HUGE benefit of social media. You can guarantee that if a NY Times best-selling author or the editor of Writers Digest posts a blog, I likely need to read it (and likely so do you).

The really neat benefit of following these two rules is that, if you establish a reputation for posting good, safe content, you transition from being a **regular user** to a **referent influencer**. Post good stuff long enough and you can graduate all the way to **expert influencer**. Also, because you are being kind and respectful to others by posting links to their sites, they likely will respond in kind.

A crucial note, here. **Take good care of your expert and referent influencers.**

As a rule, it is a good idea to make sure you eventually repost content for those who refer yours, provided the content is good.

**If your experts are gems, then your referent influencers are GOLD**. Most people will automatically make sure they take care of the experts in their networks, but the referent influencers (for reasons cited above) are probably far more important and more easily overlooked. These influencers are priceless to anyone who wants to have maximum impact with their content. Take care of these people and the return on investment is immeasurable. Conversely, alienate them and you risk losing one of your greatest assets, and the damage to your network could be immeasurable.

**A good rule of thumb is to just be good to all of those who are good to you.** At the very least, thank them. But sometimes, it requires more. If you see that someone reposting your content looks like they are trying to build a following too, and they meet Rule #1's criteria (good content), **help them**. Or don't be surprised when they move on. Networks are hard to build, and we need as much help as we can get from our social community. So if others help "raise your barn," make sure you pitch in with theirs. It is just good manners.

I might qualify, I advise being kind and reciprocating because it is the right thing to do. But, we do have to deal with reality. We only have so much time. Yes, we need to be good to as many as we can, but we need to be mindful to pay attention to those with greater reach and influence if we hope to have time left over to write great books.

**Rule 3—Always be positive and edify others (Conversation).**

When I teach social media, I tell people that ***if you have anything bad to say about someone else, only say it to people you could sue in a court of law for damages if they repeated it.*** So feel free to gripe and backbite to your lawyer, your therapist or your priest. Otherwise? Always be positive, positive, positive, and have a servant's heart.

Motivational speaker Zig Ziglar said, "You can get everything in life you want, if you will just help enough other people get what they want." If you look to how you can best serve others, your needs will be taken care of. Trust me. Promote others with a genuine intention of helping them succeed, and it will pay you back tenfold.

Why? **We like positive people.** They inspire us and lift our spirits. Have you ever had a negative person walk in the room? You can almost feel the energy drain right out the door. Saturday Night Live even did a skit on just this sort of person, and the character, Debbie Downer, soon became a slang term for anyone who feels the need to constantly share bad news and negative feelings, thereby bringing down the mood at a gathering.

But what about positive people? We can't find enough of them. People who make you smile, who make you feel encouraged and feel good about yourself are simply priceless. We like these people and want to help them to succeed...because they are the good guys.

Is it okay to post your trials and disappointments? Sure. But make it a once in a while thing, or buttress with a bright side. We know life has ups and downs and we do like to share, but always try to be positive. People gravitate to the beacons of light. We need more light. The world is full of doom and gloom and divorce and oil spills and unemployment lines. Griping about that on social media just depresses everyone around you, even if you are a political writer.

******Tip—To any political writers, you must present a solution to all the stuff you gripe about or we just feel hopeless and helpless. You might be dealing with negative issues, but that in no way permits you to be negative as well. If others walk away feeling angry and sick, they aren't likely to want to continue following you, let alone buy your book.***

This brings me to an issue of etiquette. Do not talk about sex, politics or religion unless it happens to be your platform. If you happen to be Rush Limbaugh or Bill Maher, then gripe all you like about politics. If you are a sex therapist or a romance author, then talk about sex. If you write bible study guides, then talk about religion. It is okay for you guys because you will be speaking to your target audience. If you sell Christian fiction, Christians will be buying your books, and likely not offended by you mentioning Jesus.

Everyone else? You risk ticking off and alienating at least 50% of your readers. I have had writers I had to unfollow because they're foaming at the mouth political ranting just gave me an upset stomach. The constant barrage of hate and negativity just made me attach negative feelings whenever I saw these individuals post.

That is bad, bad, bad!!!!

We want your followers to associate you with positive, happy, warm, fuzzy feelings. So leave the graphic sex talk to the porn bots. Religious and political ranting isn't good for you, but if you must indulge, relegate this for your personal social media page, and I advise you use a moniker and nothing close to the name you are branding.

Guard your reputation. People, for reasons yet unexplained, remember the negative far more than they do the positive. Probably goes back to that we are hard-wired to survive (*Note to Self...stinky oysters make you puke for three days*). You could hear of twenty restaurants that have great food, but you will **remember** the one that had a roach in the salad...and you will tell everyone you know. Our actions are the same. If you don't believe me, Google, "angry author tweets about bad review." No need to name names and rub salt in any farther. Her reputation crashed like a meteor, and she had to disappear off of social media. Now instead of being remembered for her fiction, she will be remembered for having a bad day and reacting poorly. Did she stay up all night thinking of ways to sabotage her career? No. But she failed to safeguard her most precious possession—her reputation. Please, please, please learn from her error.

Social media is a very friendly place, but always be aware that it is open to the public and it is forever (yes, even if you delete it). Thus, back-biting, griping, complaining or ranting have no place on the Internet. There is too much risk that it could just spin out of control and take your hard-earned reputation with it. Besides, negativity, by its nature is divisive. It forces others to choose sides, which is never good.

Also, remember earlier in the book we discussed the subconscious mind. By being positive, you will get in the habit of using positive goals/positive verbs, and thus run a far better chance of your message actually being heard. So when you post on your status update, always, always, always frame in the positive.

Don't forget to preorder your very own copy of my book!—(NO!!!!!!)

Preorder your very own copy of my book today!—(YES!!!!!)

In the end...BE POSITIVE. It is good for you more places than social media. Highly successful people make being positive a habit.

## Groups

Another way to build platforms on social media is to join groups. We discussed these earlier in the definition section. All three social media platforms have groups that you can join. We will discuss how to find these groups in the individual sections. Earlier I asked you to profile your reader. This is one of the reasons why.

If you write romance, your readers are 31-49 year old females who are in a relationship. What do those women do with the rest of their day when they are not reading a book? They are mothers and they love chocolate and talk about diets. They watch *Desperate Housewives*. You are a romance writer, and likely you started out as a reader first. What do you like to do other than read and write? Go to those groups.

If you're selling books on UFOs and origin theory, your fans love *Mystery Quest* and *The Nostradamus Effect*. Run searches for groups that chat about similar subjects: Atlantis, Nasca lines, Easter Island, Big Foot, Nessie, Prophesy, Apocalypse, Devil's Bible. Join those groups on MySpace, FB, and Twitter.

Get plugged into groups filled with people who like to argue about what season of X Files was best. Who is hotter, Aeryn Sun or Xena? What show was better, Stargate or Battlestar Galactica? Which is better, Star Trek or Star Wars? Who was better? Kirk or Picard? Who was right? The North or the South? What is the best romance movie of all time? *Top Gun* or *Casa Blanca*?

This is the time to use that creative brain of yours. Searching for groups of readers is not creative. Every writer trying to hawk a book will likely be running the same searches and joining the same groups. That why every one of those groups tends to be mostly writers or aspiring writers. They all want the ever elusive golden unicorn—the voracious 1-2 book a week reader. Join these other groups and I guarantee you will have better luck. Why?

You will be thinking differently than every writer out there joining reader groups. Thus, you will have a lot less competition in the pond. You are now targeting groups of people who like your topic/subject. Now, being likely one of the handful of writers, you are more apt to stand out to the group as a whole, making it easier to build a fan base comprised of potential readers. Trust me, if someone likes to spend his free time discussing Roswell on-line, this isn't a person who is too tough to convince to make the leap to reading your blog about space aliens and then to reading your book.

I also recommend joining groups of writers for inspiration, education and encouragement. But make no mistake *these groups are not the place to build your fan base*. Get out of your comfort zone and go meet your future readers. You will learn pretty soon that you have a lot in common.

Now if you lose track and spend hours on the Web chatting back and to, it is for a purpose. I see too many writers get infatuated with their on-line writing groups at the expense of other more meaningful activities. If you write Regency romance, feel free to spend three hours in the *We Love Mr. Darcy* group. Probably some writers in there, but far less than in the *Regency Romance Writers* group. You have a far better chance of connecting with readers, and that three hours chatting with people who love this time period is time well spent networking with readers and experts who might be helpful to you with research.

### Comments

**On Blogs—**Most bloggers put a lot of hard work and care into their posts. There are few things that can brighten a blogger's day more than a thoughtful comment, and there are few people we like more than those who comment regularly. Even if our stats are showing that 200 people stopped by our post that day, we still seem to have an emotional crisis if no one comments positively or even negatively. We start wondering if people just read three sentences and then hated it so much that they left as quickly as they came. Hey, we're a tad insecure.

Leaving thoughtful comments is a great way to really improve your network. Remember earlier we discussed the expert and referent influencers. A lot of them write blogs. Get on their good side and you will have great allies to help improve and expand your social graph.

### On Status Updates (MySpace, Facebook, Twitter)

Commenting on status updates (tweets) is essential for creating the camaraderie that will be beneficial in building your platform. I actually always have all three open on my computer (minimized while I work). At regular intervals, I scan all three for something that makes sense for me to comment on. If someone is having a bad day, I type a quick note of encouragement or something to make them laugh. If someone looks like they need their message reposted, I repost it. If someone met their writing goal or accomplished something they felt important enough to post about it, I take two seconds to type back "Congratulations!"

We often fall into the misguided thinking that it's the big things that matter, when in reality it is an accumulation of all the small things. Aristotle said, "We are what we repeatedly do. Excellence, therefore, is not an act, but a habit." Make it a habit to have the heart of a servant, to genuinely care and to be an encourager. The people around you need it and will appreciate it. These small acts of kindness take only minutes out of your day, and not only are they a blessing to others, but they will eventually be a huge blessing to you.

### On MySpace/Facebook

MySpace provides a way to comment on people's MySpace page and on Facebook it is called "Writing on Someone's Wall." Basically, it requires you taking a couple of minutes out of your day to write something helpful or thoughtful on someone else's page. Again, it is an opportunity for random acts of kindness focused on serving others.

Social media works best when you encourage others to support you, your work, and your cause (selling books) because they like you and are vested in you personally. When I teach social media, I see far too many writers just concerned with blitzing their announcements out to as many people as possible. Yes, social media can help your message reach thousands of people, but the techniques I am teaching help you build a sturdy platform designed to enlist the help of others. To gain the help and support of other people you need to be interested in them, help them, encourage them, and serve them. Most of the time, they will do the same and THAT is when the real magic happens.

## Rules of Social Media

Social media is a society all its own, and all societies have rules. Some we have addressed, and some we haven't. This next section is to give you pointers that will ensure that you are courteous and friendly and a good citizen of the Internet community.

The first five rules have been explained fully, so I am not going to expound.

1) Be positive.
2) Be informative.
3) Be entertaining.
4) Reciprocate.
5) Avoid taboo subjects unless they happen to be your platform (no sex, religion or politics)

Ah…but the next five, need elucidation.

6) Be smart, and be safe. Limit how much of your personal life you expose.

It is vital you show us slices of your life, but getting too familiar can just get weird. That and one must always appreciate that on-line predators do exist. They are real and dangerous and do not care if they destroy your life. In military counterintelligence it is called OPSEC or **operational security**, which basically means *think like your predators and deny them access.*

Be personable, but posting pictures of your house with visible landmarks and your street address is asking for trouble. Feel free to tweet away night and day, but don't divulge you are leaving for the day and no one is home. Don't post personal information that could be used to steal your identity like your birthday. I actually do not post pictures of my family on my open profile pages. I have a private Facebook page that only friends and family can access. I also am very unspecific about my home town and use DFW instead of naming the small country community where I live. This isn't to make you paranoid, but it is to keep you safe.

By the same token...

7) Your profiles must be open to the public.

Your Facebook, MySpace, and Twitter pages must have open access just as if they were a web site. If you really feel the need to post pictures of your kiddos and gripe about politics, do it on another page. There are few things I find more frustrating than trying to follow someone and they have their information locked down so tightly that I wonder if they are in Witness Protection.

***Here's a tip***. If you are in Witness Protection, then the feds really are not going to be happy if you are on Twitter. If you aren't in Witness Protection and you desire to build a fan base, don't have potential friends/followers jump through a bunch of hoops. We aren't that motivated. Honest. If you are that high-maintenance before we even have befriended you, then we will just move on to someone who doesn't make us solve a string of CAPTCHAs, know their last name and personal e-mail address and their favorite color to follow. Seriously? What on earth are you posting that needs that kind of security protocol?

You might be laughing, but I have run across authors on MySpace who did exactly that, and they used the *covers of their books as icons!!!!* Not only that, but they even had their MySpace e-mails locked up tighter than Fort Knox, so I couldn't even send them a message to let them know they were being annoying. So apparently these writers only wanted to sell books to their friends and family and people who knew the secret handshake. Fair enough. I won't buy their books.

Be smart and be open. Get a personal page if need be.

8) Update regularly.

If you tweet once a day or visit your MySpace page once a month, how effective do you think it will be? There is no reason you cannot visit all three sites in the course of one day. Yes, you will be building all three for reasons I will explain later. And that means *physically going and logging in*. Linking Twitter to update Facebook doesn't count. We know you are never on Facebook if we always see the Twitter icon in front of everything you post. And MySpace tattles on you, too. It says to all who view your profile the day you last logged in. How vested are MySpacers going to be if you haven't logged in in months? Your lack of interest will make them feel like unwanted stepchildren, so why would they buy your books?

It only takes a minute or two to visit the pages you use less often and check for messages, comment back to people who have posted a comment or sent a message, and do all the little things that add up to a big difference. Later we will talk about some ways to make social media manageable time wise. But think of it this way. You could own a restaurant and be an award-winning chef. Cooking can be your love. But, at the end of the day you still must allot time to market and promote or you will go out of business. Same with writers.

9) Make time to increase the size of your network.

I actually slot time to go follow so many people and increase my numbers. I carve out time to participate in groups. If you don't, your network will grow at a fraction of the rate it should to be effective. This goes back to allotting time for marketing. Allot 15 minutes per platform each week to go poach friends from the top people in your network. If you write like Sandra Brown, then go poach (befriend) as many of her fans and followers as you can. If they like her, then they may like your work. Network with readers who likely will enjoy your content.

Get creative. If you are a YA author and have teenage kids, they likely are on MySpace or Facebook anyway. Log into your account and enlist them to add friends for you while you cook dinner. That way, **when you are on social media**, you can focus on the chit-chat and networking. I actually have my husband add people for me, and that leaves me more time to comment on pages and blogs. If you have a budget for it, hire an intern or a college kid. Some of this can be delegated. Pay your kids a wage instead of an allowance. It is now a tax deduction.

10) Pick at least one platform and do it well.

With certain exceptions we will discuss, *participate on the platform you enjoy*. Don't listen to people who say MySpace is dead (it is far from dead) and switch to Facebook because it is supposedly the hot thing. If every time you get on Facebook, you want to slam your head in the door, people will *feel* that.

If you enjoy something you will spend more time on it, learn it, and are more apt to participate with the kind of regularity required for social media to be of any benefit to you in creating a platform. I recommend participating to some extent on all three because it isn't that hard to do and you will be able to expose your content to the die-hard MySpacers and Tweeters alike. Why miss reaching potential readers if you don't have to?

Now you know some basic lingo and rules of engagement, so now we are ready to get to the nitty-gritty. Yes, time to build your platform. This book is going to address the most basic aspects of building a social media network. There are a lot more nifty gadgets and ways to optimize and advertise, but that information is more advanced and for another book.

**Recap...**

By this point, you should have:

1) A brand name (name that will appear on your book)
2) A list of goals
3) An attractive headshot
4) A list of tags
5) A collection of content material—articles, reviews, photos, videos that all serve to establish your brand
6) A starting inventory of 10-15 articles (blogs) related to your topic and to serve as **strategic content**
7) Three different bios. 100 words. 300 words. 500 words.
8) A detailed profile of your reader & consumer demographic

What is the next step?

Time to dominate the domains. You are ready for Stage Two.

## Stage Two—Taking on Technology
## Dominating the Domains

By this point, you should have figured out what your brand is going to be. If you already have a website, it needs to be your brand name. If not? Time to start anew.

The first thing I recommend doing is that you need to purchase your domain name (purchase the right to use an address). I recommend www.godaddy.com to search and see if your domain name is taken or not. Type your brand into the browser (I typed in *kristenlamb*)

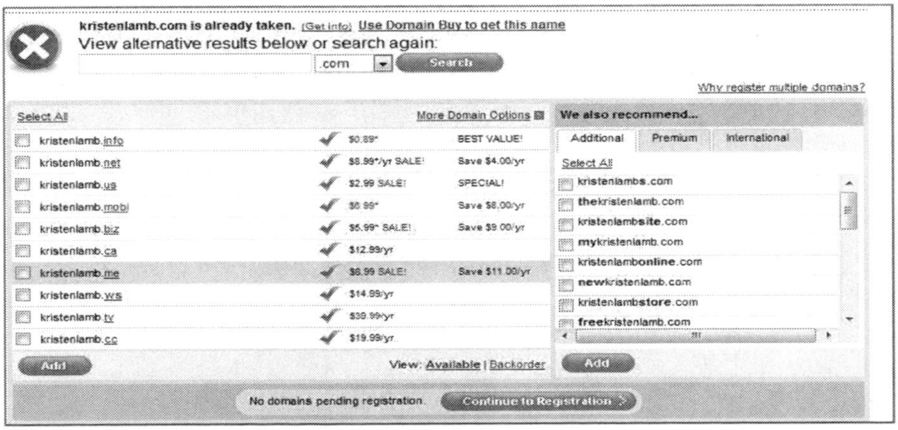

***DOMAIN NAME SEARCH IMAGE***

As you can see, Go Daddy provided me with a list of other options to choose from since www.kristenlamb.com was taken. Now, I happened to settle for www.kristenlamb.org but I could have purchased www.thekristenlamb.com or put in a bid to see if "the other Kristen Lamb" would be willing to sell her domain to me. I happen to know "the other Kristen Lamb," and she is sweet woman, but also a big media mogul. So I am thinking she will want to keep it.

This step will cost you money, but it isn't a large amount. Why I recommend this step first, is that you would be smart to try and keep your other domains as close to this as possible. Also, you will own your name and domain and that will keep others from squatting (buying it and developing it for their own agenda). If you are new, that might not be much of an issue, but there are people on the Web who will capitalize off the work of others and use your name and reputation to drive traffic to their site/business. Also, if you wait too long and happen to have a popular name, by the time you put in a bid for a domain, you could end up with something whacky.

Are you going to use this domain to build a web site? Eventually, but that isn't necessary right now.

Your job now is to dominate the domains, so all you are going to do is sign up for membership and plug in the bios you used earlier. That is it. Leave finding friends and importing address books for later. We want your friends and family and fans to see pages that are complete and spiffy, not a chaotic construction zone.

***"NO PHOTO" IMAGE***

If you take off too quickly, this is what visitors will see, a silhouette and no followers and no content. Not the image you want to start building your platform.

Wait. Your patience will pay off.

**If you are techno-savvy, follow this next set of instructions. For those who are new or unsure, just begin with the next section.**

If you are good at following digital cues, then I recommend signing up for the Wordpress, MySpace, Facebook, and Twitter accounts in one straight shot. Just get an account opened and put in your bio if you want to. Then you will start fleshing out beginning you're your Wordpress blog and then MySpace as your **first social media platform.** When you sign up for a MySpace page, paste your bio information in the box noted as **General**. You will be using the **About Me** box for something different.

Then sign up for a Facebook page and a Twitter account. All three should be fairly simple to do. These sites have spent a lot of money making it easy to at least do the basics. On Facebook, you will find there is a place to put personal information like cell phone numbers. I strongly caution against this. Maybe I am paranoid, but we do have to appreciate that hackers exist. Facebook has privacy settings that allow you to only display your home address and cell number to friends and family, but I think it is too risky. If someone needs your cell number, let them e-mail and ask for it. Why risk making a mistake and incorrectly setting the security protocols? I am often multi-tasking so I choose to minimize my chances of royally goofing up. I recommend you do the same. Not only that, but Facebook has been in quite a lot of hot water these days over privacy violations. Be careful who you trust with information. If you wouldn't put it on a wide-open web site, keep it to yourself or do so at your own risk.

Twitter will be the simplest of all three and I recommend doing it last. When it asks for your web page, you will eventually use your MySpace domain (in addition to a web page if you want). **Right now, just make sure all the accounts are active and ready for you to begin building**.

Resist the urge to import address books and start looking for friends. You can fill out basic information, claim a domain, and fill in the bio, and maybe upload a picture if you are more tech savvy. But then stop there at the ground floor. You have some work to do before it is time to invite others to your sites.

## Go Get Your Free Blog on Wordpress

So you purchased your domain name (or have plans to do so soon). Your next step is to **go and sign up for a free blog on a hosted site**. I recommend Wordpress for a number of reasons. Wordpress is easy to sign up for. It is free and simple to use. Pretty much just read and follow directions, but I will give an overview. The beauty of Wordpress is you don't have to know HTML or even a lot about computers. It is a technophobe's best friend. Wordpress takes care of all the gadgetry so all you have to worry about is the content.

If you are more computer savvy or you learn easily, you can eventually graduate to a level of Wordpress that allows you to install software and have more control over the look and interactivity of your blog. For most of us, though, it works just great to have Wordpress do all the geek stuff for us.

Wordpress has beautiful free backgrounds and is easy to optimize (make your content easy to locate and share). Also, if later in the game you decide you want more (ability to have streaming video, for example) you can purchase additional bells and whistles at a reasonable price. Wordpress employs a very good SPAM filtering system known as Askimet and aggregates your hits for you so you can easily see how many people are visiting your blog per day/week/month/year. Later, this data will help you be able to more easily see content that is more popular and allow you to adjust your content accordingly. A

Wordpress blog is also easily transitioned to a Wordpress optimized site for later on when you are ready to get your own web site.

Wordpress, during your registration process, will ask you to choose a domain name. **Make sure when you choose a domain name, you choose your brand name or as close as possible**. I chose KristenLambTX (also my Twitter handle) because that will appear in the URL. (MySpace calls this your **unique name** but it is the same as a domain and *cannot be changed.*)

1) Go to Wordpress.com and click to sign up for a free blog. Fill out the first page of user information.
2) The next page is information about your blog's domain and name. **The name can be changed, the domain cannot.**
3) In the tabs above, you will see the first tab is **My Account**. Place your cursor over this and a drop-down menu will appear. **Select Edit Profile**. Feel free to fill in all the bio information. Your bios should be the same across all platforms. This consistency will help your search rating (where you sit on a Google search, for instance). You will soon see why I had you write bios of different lengths. Wordpress uses Gravatar and the bio is all of two sentences in length.
4) Click **New Post** and you will be forwarded to a page that says *Add New Post,* with two windows beneath. One is long and narrow and sits right above one that is large and fat.
5) Select one of the blogs you wrote (strategic content).
6) Paste the body into the window.
7) Type the title in the title window.
8) Click the box on the right-hand side that says **Save Draft**.
9) In the long box below the body, you will see a box for an **excerpt**. I recommend you cut a sentence or two

from the body of your article that sums up the content and paste in here.

10) Hit **Save Draft**.
11) To the right-hand side, you will see a box that says **Categories. Click +Add New Categories.** Type in the topic you will be blogging about (UFOs, for instance).
12) Make sure you check this box or Wordpress will automatically slot your blog into **Uncategorized.**
13) Below the **Categories** box, you will see **Post Tags.** This is where you add in all the nifty tags you came up with earlier (the ones related to your blog post). Add in any more tags that might make your blog easier to find in a search.
14) Click **Save Draft.**
15) Now hit **Preview** and scan your post for any errors.
16) If it is error-free, then hit the **Edit Post** option at the top and this will take you back to the page you were using a moment ago. Hit **Publish** and *voila!* You have a blog.
17) Go back to your **dashboard.**
18) Your blog isn't that pretty yet, and we don't want to start driving traffic here. So now look to the left side and there is a menu to the left. About half-way down you will see **Appearance**. Click this and you will see all kinds of options for how your blog can look.
19) Scan the different options. Look for ones that are eye-catching, attractive, but most of all, **easy to read!**
20) Preview how they will look by hitting....you guessed it! **Preview.** This will help you make sure you have just the right background by giving you a peek at what visitors to your site will see.
21) Click **Activate** to activate your new background.
22) Click **Visit Site** to see how it looks with your content.
23) If you are happy with it, leave it as is. Or, click **Appearance** and try another skin until you find what you like.

24) You can post more blogs, but it isn't necessary. Very little traffic will be heading your direction yet.
25) **To insert attractive hyperlinks**, go to your dashboard. Highlight the word you desire to attach a link to. For instance, I might not want to break the flow of a blog on writing. What I can do is highlight the word **Writers Digest Magazine**. An icon that looks like a chain link will appear. Click the chain link and a pop-up menu will appear where there will be a prompt to insert the link www.writersdigest.com. Then, and this next step is important, I will be given a number of selections of how I want this new page to open up. Select **open link in new window.** We want to serve our readership by giving them digital breadcrumbs leading to other noteworthy sites, but not at the expense of our own page. Doing it this way will just make certain selected words change colors as a visual cue that these words are clickable hyperlinks.
26) Log out. You have more work to do!

## MySpace

MySpace? Yes, when it comes to social media, you begin with MySpace. In my opinion, *all new writers must have a MySpace page.* If it were up to me, all writers would have a MySpace page. Why? Because MySpace pays me every time I say that. Just kidding.

Many of you might already be on MySpace. You don't have to start all over with another page if you don't want to. Just change your username. I had texaswriterchik as the URL, but also had 600 friends at the time and didn't desire to start all over. I just changed my user name to Kristen Lamb.

If you are new to MySpace or you don't mind starting over, then we have work to do. If you build a brand new MySpace page, worry about friends later. You have a highly optimized network to build. I want your whole network in place before we worry about making any connections.

If you already have an existing MySpace, you can either start a new one or just change your username. Ideally, I want you to have as many domains with your brand name, but if you already have 900 friends, then just stay right where you are, but make sure your user name is your brand name.

If you desire to start a new MySpace for a pen name, just make sure to tell friends and family what you are up to so they know the friend request is from you.

**Getting started...**

I hear a lot of people tell me that MySpace is dead. Well, 60 million active users says otherwise (those were the stats as of March 2010). But, we will talk more about that later. MySpace should be used for one singular reason. It works far better than a static web site. Yeah, that domain name you purchased earlier? You don't have to use it until you really need to.

(And MySpace pages are free...all right, two reasons.)

How many of you (especially new authors) have $10,000 to go drop on a fancy, interactive web page? How about $5000? What about $500?

Okay.

How many of you really want to pay a webmaster every single time you want information changed? How many of you, even if it is free, want to be chained to a webmaster every time you desire to post something new? Static web sites have a lot of disadvantages and most writers have to make it to a certain point before the expense is really justified. I have found that my MySpace is far easier to check the e-mail messages and modify content.

http://www.myspace.com/texaswriterchik

Texaswriterchik. Yep. See, I keep my mistakes around so you guys can learn. Yet, even though I made a critical error (MySpace was my first social media experience), I have had over 30,000 hits to this page alone. Not bad numbers in the greater scheme of things. Not only that, but until recently, when you

Googled my name, my MySpace was at the top of the list, *above* www.kristenlamb.org.

My methods will help you dominate the searches. This positioning might change slightly day to day because the web is always in flux, but, by posting contiguous bios and content that uses the same tags, you can dominate the first page of all searches. Google *Kristen Lamb* and you will likely find MY MySpace page, web site, LinkedIn, Twitter, and Flikr accounts all on that first page and often in those first slots.

The "other Kristen Lamb" is in media. I basically booted her from all of the top slots simply because I constructed all of my sites in ways that used the search engines to my advantage. This is what I am helping you do as well.

Back to the static site. Unless you pay more money, most of the time, you are stuck with some pretty boring templates. Even though I had a good friend create me a very impressive header, I still ended up hiring a firm to jazz up my sites even more and optimize them as well. Unless you happen to know how to build web sites, optimizing them is no walk in the park (inserting widgets and RSS feeds). There is a time to outsource, and we will talk more about that later.

Also, a static website is harder to boost in the search engine rankings unless you blog off it because the more you add content, the higher you rank in a search. That, in my opinion, is far easier to do with a MySpace page. I can add pictures, update my status, add a link to my blog (we will go over that later) almost daily and without headache. At the end of the day, I want to be able to reach out to all my potential fans and communicate with them on the platform of THEIR choice, not mine.

Thus, we are going to start with building a MySpace page because it is easy, and you will soon see that it simply looks better than most beginner web pages. It is simple to link to all other social media platforms, and is extremely searchable. And best of all, it is already integrated into a network of millions of MySpace users.

MySpace likely will be the most time-intensive of all the pages you will build so we will begin here first. Once you have done it, though, it is very simple.

**Step One—Sign Up for a MySpace Account (if you haven't already)**

Remember, your **unique name and domain name are the same thing.** Fill out all the information and you can even paste your 500 word bio in your **General Section. Leave the About Me blank.** That will be used for something else.

This part should be relatively easy. It is just filling out boxes that tell about you: hobbies, movies you like music, etc. Feel free to fill out as much or as little as you like. I recommend at least writing *something.* Keep the sections brief and the **General (bio)** no more than 500 words. Feel free to put your blog's URL in the **General** section below your bio, but I chose to place it in my Music section, since I felt that was a better use of that slot than all the bands I have ever liked.

**Step Two—Search Backgrounds**

You now have a MySpace account, but we need to do some ground work before we worry about making any friends. By now I hope you have an idea of the image you wish to portray, so the first thing we are going to do is search background wallpaper for you page. No plain white pages. Who wants to look at nothing?

Now? Yes. No time like the present.

I recommend these two sites for the best backgrounds and they are FREE:

http://www.freecodesource.com/

http://myspaceoryours.net/

You will probably have to close a lot of pop-ups, but other than that slight annoyance I have never had any trouble with either of these pages. They offer free backgrounds with amazing artwork. The hard part is just sitting and looking through all of them and trying to figure out which one you like the best.

***MYSPACE OR YOURS? IMAGE***

These sites divide their backgrounds by topic from abstract art to urban. Or you can look at the top ranked pages and see what other people liked the best. This is really the fun part of MySpace and why I like it so much. MySpace allows me to express my creative side far better than any other social media site.

*****While you are here, you may want to search for Twitter backgrounds as well. On the opening page of the MySpace or Yours site, you will see a link to Tweetygotback.com. This site has some beautiful Twitter backgrounds and many match the MySpace pages, so you can coordinate. Twitter works a bit differently, than MySpace. There will be no code to save, so just bookmark the page where you find the background you want so you can find it later.***

**Step Three—Save at Least 5 MySpace Backgrounds in a Word Folder**

I recommend that you spend time looking through the backgrounds and ***save at least five different backgrounds***. Yes, you read correctly. I save them as Word Documents in a *MySpace Background Art* folder. That way, you can change the background and keep your page with a fresh look without having to spend a day sifting through backgrounds. I try to change my page backgrounds at least seasonally. You don't have to, but it is a good way to drive traffic to your page. Every time you add or change anything to your page, it notifies all those following you in their stream (update on what everyone is doing in your network).

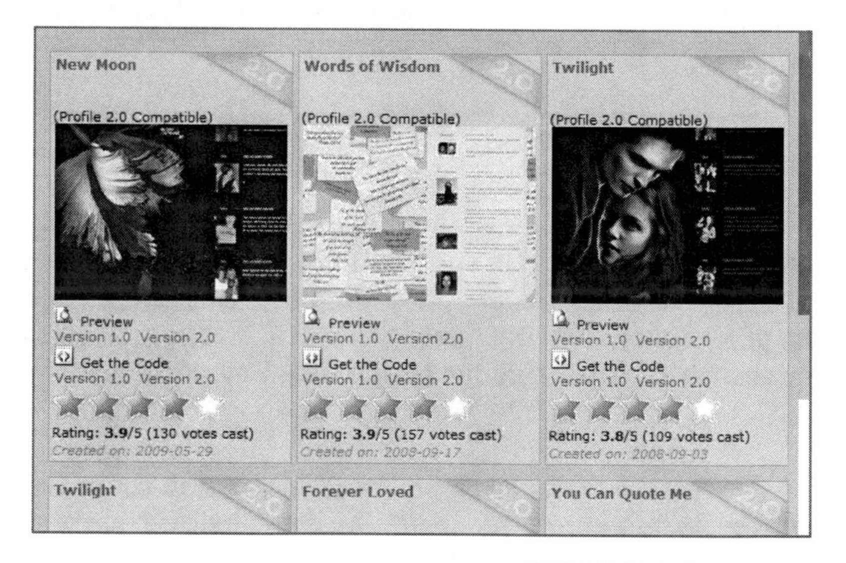


***MYSPACE BACKGROUNDS IMAGE***

First, you see Profile 1.0 and Profile 2.0. The only difference that makes is how the information is laid out. Click on either and you should be able to see the difference and choose which you prefer. I use 1.0 for my page, but when I built Bob Mayer's page, the best background for his genre only came in 2.0. Most of the sites will let you get a preview of both versions.

Then you might notice stars beneath each of the images. You guessed it. That is a ranking by users as to how the page looks. It won't take long before you realize that not all pages are created equally. But there is some amazing art that I strongly recommend you take full advantage of. Why not? All it costs you is time and a little effort.

You will be able to see a live preview. This is a way for you to see what the background will look like BEFORE you change anything. Then you will see a button that says *Get the Code* and it will give you a choice between 1.0 and 2.0.

When you select either, a window will pop up that has some scribbly-looking stuff that might resemble this:

<style type="text/css">.fcsS{}</style>

This is the HTML code that will generate your background. Highlight this area then paste into a new Word document then label what it is. For instance, "Cool Dragon Background." Locate and save a handful of these so you can save time when you desire to change your background. All the hard work will already be done.

Want a custom layout? I recommend www.webfetti.com for ease of use without the spyware/adware problems.

Found the profile background you want to use? Now you are ready to get building.

**Step Four—Building Your Page**

1) Log into your MySpace account
2) You will see a menu right below the MySpace icon and right above *Hello, Your Name Here*!

***MYSPACE HEADER IMAGE***

3) Go to Profile and a drop-down menu will appear
4) Select *Edit Profile*

5) Copy and Paste your background code into the **About Me** section. You should have pasted your bio into the **General** section, but, if you didn't, no need to worry. Just make sure you scroll down below your bio and **then** paste the code. If you mix the two up, it will look like what happens when something doesn't go through the Star Trek transporter correctly. A mish-mash jumble of "*Ew*!"
6) Select the **Save Changes** to, you guessed it, save your changes.
7) Select **Preview Profile** to see what it looks like. Maybe that background was far cooler in theory than on your page. Is it hard to read? Is it eye-catching? Once you are satisfied, just follow up with a last **Save Changes** and you are ready to upload your profile picture.

### Step Five—Loading a Profile Picture

1) Click **Home** and go back to your dashboard. Below the silhouette (a future picture of you) you will see a number of options. Click Upload.

***MYSPACE PHOTO UPLOADER IMAGE 1***

2) Upload your profile picture into a photo album.

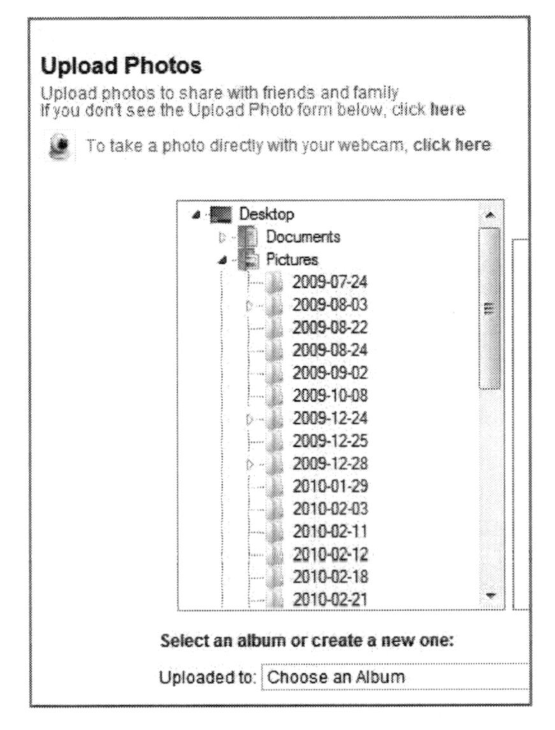


***MYSPACE PHOTO UPLOADER IMAGE 2***

When you hit the **Upload** button, MySpace will automatically display a dropdown menu of all of the folders it has found on your computer. Select the folder where your profile picture is saved then click **Upload.**

3) You will have to put your pictures in an Album. *Profile Pictures* is adequate.
4) Once the images upload, you will see your pictures and blank boxes where you can choose to add captions and tags. You can save that for later by choosing **Skip this Step** at the bottom (If you do put in captions and tags, then choose **Done Editing**).

***MYPACE PHOTO ULOADER IMAGE 3***

5) Next you will see your photos.
6) Double-click on the photo you wish to be your main profile image, and you will see that the picture enlarges. Above your photo, a menu will appear. Your first choice (going left to right) is **Set as Default.** Click this.
7) A box will appear asking if you are sure you wish this image to be used as your default image. If you do, then click **Okay.**
8) You will see the new image appear where a silhouette once was.
9) Go look at your actual MySpace profile. You can do this simply by moving your cursor over your profile picture and double-clicking. See what it looks like all put together and what, if anything, needs changing.

### Step Six—Adding Music

1) Go back to your dashboard (click **Home**). Remember that menu above the *Hello, Your Name Here!*?
2) There is a button that is labeled **Music.** Click this.

3) In the top right corner, you will see a white search box. This search box is internal to MySpace. If a musician or band has music loaded on MySpace, you should find it. Type in the name of the band or musician.
4) If your choice is represented, it will appear in a list of songs. You can see if the song is by the actual artist or someone else.

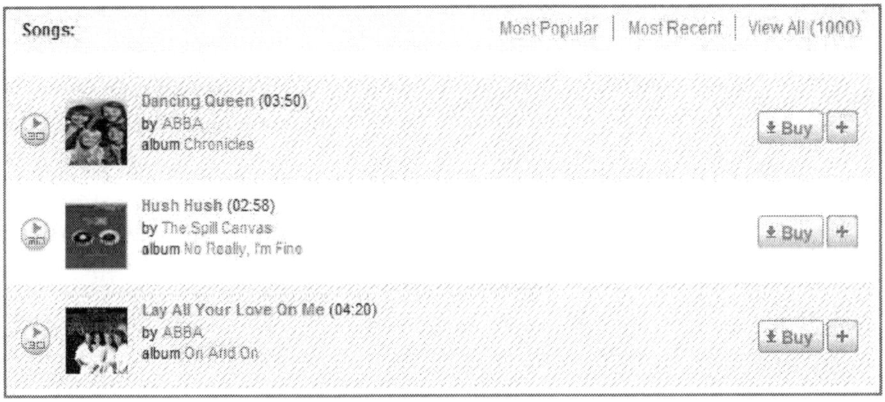

***MYSPACE MUSIC IMAGE***

5) You can play 30 seconds to make sure this is the version you would like on your page.
6) Hit the + sign to Add to your playlist
7) A bubble will appear.

***MYSPACE "ADD TO PLAYLIST" IMAGE***

8) In the blank box, just write "My Songs" or something to that effect. You only want to load one song. When you click **Okay** the song will automatically be added to your page.
9) Go back to your dashboard (Click **Home). Click Edit Profile.**

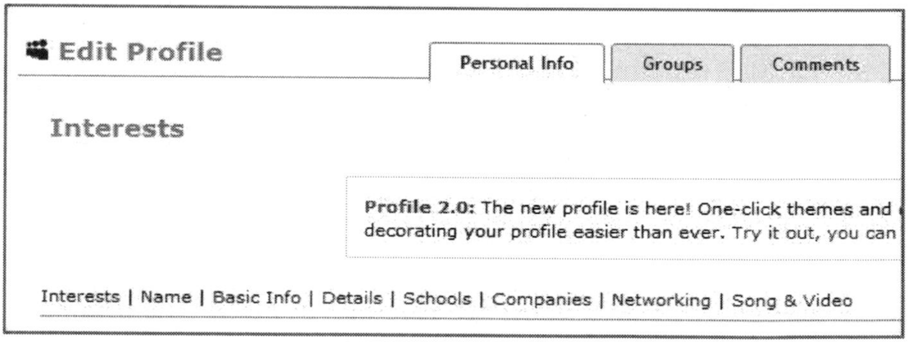

***MYSPACE "EDIT PROFILE" IMAGE***

You will see there is a tab called **Song & Video.** Click this and it will bring up a page that has the settings for your music and video additions. The first on the list says **Profile Song: Change auto-play settings in**

**Account Settings (in blue). Click this blue part** and this will bring up a list that will allow you to choose if you want your music to start immediately when someone opens your page (recommended).

******Tips—Avoid creating long playlists. First of all, who is realistically going to spend an hour on your page listening to every song you have liked since high school? The more songs you add, the more data has to be transferred when someone goes to your page. Too much data means people will grow old waiting for your site to load. Avoid adding additional music, graphics and slide shows. These extra gimmicks just slow down load time and frustrate anyone trying to look at your page.***

All your MySpace page needs is your information, a bio, your picture, and a song. Later, I recommend adding a link to your website and your blog in the **General** section with your bio or, if it doesn't stand out, use another section like **Music** or **Movies**.

Additionally, choose music wisely. Most people expect a song to load on a MySpace page, so you actually seem odd if you don't have music. By the same token, Norwegian Death Metal might not be a good choice (unless you write Horror). Choose music that goes with your brand or theme. Back when I built Bob's MySpace, I chose Nickelback's *If Today was Your Last Day* because Bob writes books on being the best and being a winner and striving for the excellent life. The song went nicely with his theme.

**Step Seven—Checking Your Settings**

1) Go back to your dashboard (click **Home**) and click **MyAccount**.
2) You will see an **Account Settings** page appear. Go to **Privacy**.

a) Uncheck that your birthday be shown to friends. If your page is open to the world this is just asking for trouble.
b) Make your profile visible to everyone. If you wouldn't want your mother to see what you post, then you probably shouldn't post it.
c) Scan down the list, making sure your content is visible to everyone. The only exception I think is fair is that you make your photos visible by only your friends. Other than that, this is essentially a Web page. Businesses do not hide their Web pages from the public and neither should you. Businesses will, however, make visitors sign up as members to access other parts of their sites, so withholding open access to your photos is fair. Hit **Save Changes.**

3) Hit **Sync** and follow instructions to Sync with Twitter (if you already have a Twitter profile. If not, you will need to do this later). This allows your status update to be continually updated from Twitter (in addition to updating from your MySpace Dashboard). You can always unsync later if this is not useful for you. Hit **Save Changes.**
4) Go to **Notifications**. I would uncheck these or it can overwhelm your e-mail. Remember, you will not be a casual MySpacer who has 20 friends and a closed world. This could make you crazy if you get an e-mail every time someone breathes on MySpace. Hit **Save Changes.**

**Step Eight—Find Friends (I recommend doing this later.)**

I suggest you build all your pages first (you still have Facebook and Twitter to go), but it is up to you. If you really want to, friend away! At least now when people visit your site, you will have something for them to see (including a link to your Wordpress blog).

Once you decide you are ready to begin looking for friends, I recommend using the search box on your dashboard page to search for people you know on MySpace. But, an easier way is to simply look at that top menu (the one that has **Home** and **Music**) and you will see **Friends**. This drop-down menu will give you a variety of options to locate people you know on MySpace. Just follow the cues and start requesting friends.

I hope all of you befriend me, at least, and even my top friends (and their top friends). It is a good place to start so at least you have some friends on your MySpace. Besides, the beauty of social media is that we all work together!

http://www.myspace.com/texaswriterchik

I also recommend poaching. If a big name author has a MySpace page, I suggest adding her friends to your friends. This same menu has one last tab called **More. Click More** and the drop-down menu will show you more ways to enjoy the MySpace experience. Roughly the fifth item down is **Groups**. Click **Groups** and follow the cues and you can start searching for all those wonderful groups we discussed earlier.

When you make a friend, look at their lists of Top Friends. If you like them, the odds are you will probably like their friends as well. Add them too. I make it a habit to add so many people per week. When you add friends, MySpace will ask if you desire to see their Status Updates. YES!!! This is how you can comment and interact.

Tera Lynn Childs Today's special delivery (for a top secret project to be announced soon). http://twitpic.com/1lkod0
Posted 16 minutes ago from Twitter
view more | comment

Geri Ahearn posted a new bulletin: 5 Star REVIEW~"PENNY & RIO:The Locked Doghouse Mystery"
Posted 18 minutes ago

Robert Farley No-cloud, blue-sky sunny day.
Posted 29 minutes ago from Mobile
view more | comment

Smoking Poet The Smoking Poet is putting on the dog!
Posted 30 minutes ago from Twitter
view more | comment

***MYSPACE STATUS UPDATES IMAGE***

Notice in this image that we can immediately see who is updating from Twitter or even a cell phone. MySpace gives me an opportunity to comment and get a dialogue going.

I recommend at least once a day going to your dashboard and scanning the stream. Comment on one or two posts if possible. It shows you are present, vested, and paying attention. It is invaluable for creating relationships that will translate into a solid platform and fan base. It also will score big points with a search engine.

Whenever you update your Wordpress blog, tell everyone in your MySpace network in your Status Update.

I advise that you get all your social media sites built before you worry about friends and groups. Once you feel you are ready, refer to your Reader/Consumer Profile (What do your readers/consumers do in the day other than read?). Use this profile as a guide to help you join the appropriate groups and begin socializing with other MySpacers who love your topic(s).

******Note—If you write YA, MySpace is a HUGE hangout for teenagers. Other genres have more leeway, but authors targeting a teen audience would be wise to have a solid MySpace presence. Fish where your fish are, and your teeny-bopper fish are on The Space.***

So now you are an official blogger and should have an active Wordpress blog and a fully operational MySpace page. Your MySpace page should have the link to your Wordpress blog embedded somewhere in the body. As I stated earlier, I love a lot of music, but I chose to put the address to my web page and my most active blog in this section instead of a long list of favorite bands. There might be better ways, but this is the easiest.

Whenever I post a new Warrior Writers blog, I just post the announcement and the link in my Status Update. You could build a really great platform with just Wordpress and MySpace, but we need to sally forth. There are millions of users who prefer Facebook and Twitter. By the end of all this, you will extend your digital reach to all of them.

## Facebook

Facebook is actually very easy to use, and if you braved it through building your own MySpace page, I think you will really be shocked how effortless Facebook is. I truly believe that it is this fundamental simplicity that has made Facebook probably the most attractive social media platform for all age groups, young and old alike. There is no need to search code for a background or embed music, so there is no wonder it is a very appealing platform for the more technologically challenged.

This section is going to cover a lot of information. Facebook is simple to use, but it has a lot to offer, so I want you to understand some concepts before we go wild filling out profiles. We are going to discuss the fan page vs. the profile page, then what they do and why they are different, THEN I will walk you through constructing both.

## Before We Begin—Profile or Fan Page?

Well, here is a real sticky wicket. There has been a lot of drama associated with having fan pages, namely because if you do it wrong, it is a formula to land in therapy feeling unwanted and unloved. Who wants to build a fan page and only your mother become a fan? No one.

I preach all the time about writers, even new writers, portraying that they are professionals who take their writing seriously as a business. In fact, I probably have said it quite a lot even in this book. Yet, there is a place and a time for everything. The underlying theme of this book is to make social media simple & effective. Fan pages are no different.

**FB profiles and FB fan pages** serve different purposes. This book is going to teach you how to build two different presences on FB. I will explain the differences between the two, but just to be clear, you need BOTH. We are going to expect success, so my recommendation is that EVERYONE build a Facebook profile page AND a fan page.

Go fetch a paper bag and do all of your hyperventilating now. I am here to save you time. If you expect success, my method makes sense. **No one said you had to publish your fan page, yet.** So all you newbies who don't even know what genre you want to write yet, have no fear.

## Why build a fan page right away?

Because I said so, ha! Kidding.

Remember when you built your MySpace I recommended waiting on friends? We are going by the same principle. While you are building your regular Facebook presence, you will be able to add content to your fan page as it comes available—logo, pictures, links you find interesting, video. You will also be able to get your blog established and gather content for your *Notes* and your *Discussions* section.

Building your page ahead of time keeps you from having to kill yourself. As you build your number of followers on

Facebook, you can be systematically adding content to your fan page so you have something for potential fans to see once you decide to launch and send out an official invitation.

**During the construction phase...**

No one needs to see your fan page but you. When the time is right, the goal is to invite your friends/potential fans to a fan page that already has content, and appears established. It will look far more planned and professional, thus diminishing the weirdness factor.

**No time like the present...**

Why I recommend doing this now is so you have your entire social media platform built and linked together under the same brand name with the same theme, the same content, and the same tags. I want you to own as many domains as possible. If you command the fan page name now, then when you are ready to go live with your fan page, all you have to do is do a bulk invite and publish your page.

******Note—You will not own your fan page name until you have 25 fans. I wouldn't worry about someone else getting your name ahead of you. Facebook takes great strides to make sure that you have the right to your name. So, I can't go in and make a Heidi Klum fan page unless I happen to be Heidi Klum or can prove I am her authorized representative. Facebook will delete the page if I break this rule.***

***But, if you have a common name or you feel it is better to be safe than sorry, then all you need are 25 fans, and the name is yours. I recommend communicating your plans to your friends and family and elicit their help. They will understand. You can go live just long enough to get 25 fans and then hide the page until you are ready to launch.***

I don't believe in One Size Fits all, and neither does Facebook. Though I recommend every writer **build** a fan page, when you **launch** this page is going to depend a lot on your goals, your existing network, and where you are professionally.

It is ideal to build the fan page or pages off the main profile page, so you will need a profile page first.

Facebook is actually going to be different than MySpace. I recommend you sign up under your real name even if you desire to use a pen name (or go ahead and keep your page that is under your real name—we will put it to work). That will make it easy for you to add friends and family to your network and build a following that you will eventually direct to your **fan page using your pen name**. We want to be able to take advantage of those close social ranks—friends, people we knew in high school, people from work. These people are going to be your most likely readers and fans because they *know you.*

This is GREAT news for the person who has been on Facebook since college and has 900 friends and was dreading having to start another page. You don't have to. We are going to use that network you have already built to your favor. In fact, fan pages are a great way to start separating your identities.

I had one of my students, Pam Laux, who did this exact thing. She was already an established FB aficionado, but everyone knew her as Pam Laux Real Estate Agent. She wanted to separate her professional/personal identity from her Pam Laux Author identity. A fan page made it very easy and she has the added advantage of being able to tap into those intimate connections.

There are others of you who have never been on Facebook, and who feel more comfortable just starting off with the pen name from the get-go. That will work, too. If you have decided that Facebook will only be for networking to build your platform and you only want to use your pen name, then go ahead and get a Facebook page using your pen name. *Just make sure when you eventually send out friend out friend requests, you alert your friends and family.*

Samuel Clements probably had to do the same thing with his own family. *Mark Twain who?*

When you build the fan page off your main profile, then you will just have to decide when the time is right to do the unveiling by going live and sending out invites. Published authors will likely go live with their fan pages far sooner. Why? Easier to hit the Golden 300, as I like to call it. I advise anyone of Facebook to wait until they hit 300 followers before sending out an invitation for a fan page. You are far more likely to hit the 25 mark to keep your name, and you are fishing from a far fuller pond. We are expecting success. If you are writing and selling and teaching, etc. you may not have as much free time to build this fan page once success comes your way. A fan page will take all of 15 minutes to build the basic structure then you can take your time fleshing it out.

If you are a published author, but you are totally new to Facebook, building your fan page but waiting to publish it will save you time, help you put the best foot forward and will spare you the embarrassment of being a well-known author with 28 fans. We are setting up your platform for peak performance, and your time is valuable. I am going to show you a way to get a large fan following almost instantly.

If you are a published author who already has a large fan following (in other places), the Facebook profile page will be very valuable to directing followers to your fan page by **using the main profile as the hub or traffic cop directing fans to the appropriate fan page**. I recommend that you embed a FB widget on your static web page (once you are finished building everything) so that readers can connect with you on all the platforms. We will talk more about this later. But, if you aren't on FB already, a lot of your readers likely are. So embed the widget and you should hit 300 super quick. From the profile page, then you can direct to your fan page. Or, if you are a big enough author, just direct your readers *directly to the fan page/pages.*

**Pages? Yes, there are instances where more than one fan page makes sense.**

## Multiple Fan Pages?

Once you find out how easy it is to build a fan page, you might be tempted to get crazy. So let's go over some scenarios where multiple fan pages make sense so you recognize a good idea from a crazy idea far sooner.

**Good Ideas...**

**If you have successfully published multiple distinctive series**, then multiple fan pages could be very helpful and a good way of managing a lot of titles, especially if these titles have a large fan following. For instance, Bob Mayer has his Area 51 series, his Area 51 Atlantis series, and his Dave Riley books. These different series make sense to have their own fan pages in that it would make it simpler for Bob to make regular announcements that are specific to each individual series' fan base. For instance, if Bob releases one of his backlist on e-book or if Hollywood decides to make *Area 51—Nosveratu* into a movie, then fans of that series would be alerted via his fan page.

Now *Bob Mayer Profile Page* can act as a hub, giving fans a choice of which of his series they want to follow.

**If you are successfully published using more than one pen name,** then multiple fan pages could be a good idea. Continuing with my example of Bob Mayer, Bob could also decide to merely have fan pages for each pen name/collaboration—Bob Mayer, Robert Doherty, Greg Donegan, Bob Mayer & Jennie Crusie. Fans who read Bob's romance co-authored with Jennifer Crusie are probably different than the fans who like his Green Beret books or even his Area 51 series. The fan page is a great way that Bob could speak to individual fan bases, if he found that would be beneficial.

**If you are successfully published in more than one genre,** then you could have a fan page for each genre. For instance, Bob has hit the best-selling lists in multiple genres. He could just keep it real simple and have a fan page for *Bob Mayer Fiction* and *Bob Mayer Non-Fiction*.

Please keep in mind, that Bob Mayer is a great example for this section because he is a statistical anomaly who meets all the criteria—multiple successfully published series, genres, and pen names. That is highly, highly unusual. Bob has 42 titles and has sold millions of books and thus has different problems than us mere mortals. When we sell millions of books, we, too, will have the need for these kinds of fan pages because we will be managing lots of titles with lots and lots of fans. But, by the same token, we likely will have the funds to hire a firm to manage all this stuff too.

In the end, even a big author like Bob Mayer will have to look at what makes sense from a business standpoint. A regular Facebook profile page alone is too limiting for someone who is a NY Times best-selling author. Bob needs the interactivity of the regular Facebook page, but he also needs a place to post regular announcements about his books, his workshops, and his blogs.

Bob, like any author in this unusual situation, will have to ask some important questions. Would it be too confusing to have everything under the banner of one fan page? Would it make sense to break fan pages into individual series? Genres? Pen names? **If an author has multiple fan pages it must be to *simplify and focus.*** If creating more than one page doesn't make life easier, then forget about it and stick with just one.

My recommendation is that you start with one and see how it goes. The way we break people out of their patterns is by introducing new, but similar things. Try to buy a book on Amazon, and you will always get a list of other suggestions. Bob may have die-hard fans of his Green Beret works that could be gently guided to being Atlantis fans.

If the overlap of genres and pen names, etc. gets confusing, then make another page. For the most part, at least in my experience, people just pay attention to their interests. Bob could have a singular Bob Mayer fan page and if I like Green Berets, I will pay attention to those posts and really just ignore the rest unless something catches my eye.

**If you are published and want to promote a special event.**

Yes. Corporations use these types of temporary fan pages all the time. Frequently, they even have a time limit, and this finite appeal can help drive traffic to the appropriate page. For instance, Who Dares Wins Publishing can create fan pages for the new books and authors and drive visitors/fan to the appropriate author fan page. There can be promotions and discounts to help launch a book, but then the page is pulled after a certain window.

If you have a new book coming out, then you can create a fan page to just promote, promote, promote, but make sure to drive all those fans to your main author fan page before you pull it down. This can help you really focus your energy behind the new book before it hits shelves to give it as much momentum as possible.

**Crazy Ideas...**

**Should I create a fan page for my unpublished manuscript?**

No.

Fan pages for unpublished works fall into the same problems we discussed earlier about Twitter identities. You have no control over if/when that work will be published and no control over title. Spending months or years building a fan page for a novel that might not ever see print is just a bad use of time.

Also, even if you were successfully published, you would be back at ground zero for the next book's platform.

**If I am a published author, should I have a fan page for my individual books?**

Again, no.

That is just a formula to be spread far too thinly (again, think multiple Twitter identities). Unless you are willing to outsource, this will be a nightmare to maintain. The only exception would be a temporary fan page to generate excitement and momentum for an upcoming book.

**Should I make a fan page for my characters?**

No. This goes back to making Twitter identities for your characters. If you are unpublished, you won't have control over the characters. A name could change, a character could change. Also, if you don't write any more books with those characters, then that is a lot of energy you could have put toward branding your name. By the time any on us get to be big enough that people will care that much about our characters (Edward, Bella, Harry Potter) we will have entire marketing firms doing that stuff so we can write. Focus on branding you.

**Profile vs. Fan Page**

**What is the difference?**

Fan Pages:

1) Are for promoting a real person, a business, or an organization, whereas profiles are only for promoting a person.
2) The information posted is unidirectional and lacks the two-way dialogue of a regular Facebook profile page.
3) Fan pages have the ability to exceed the 5000 person limit.
4) A fan page can have multiple administrators, whereas a profile can only have one.
5) Has the ability to send out bulk messages to ALL followers (Profiles are limited to 20 and Groups limited to 5000).

6) Those receiving the messages from fan pages cannot reply. They can only opt out of receiving them.
7) Followers (fans) are automatically approved, so this takes out the additional step of having to approve everyone who sends a friend request.
8) FB fan pages can provide detailed metrics broken down by demographic. Who is viewing your fan page and how are they viewing it? Profile pages have no aggregate information.
9) Fan pages cannot restrict visibility (except based on age), whereas profile pages have all kinds of privacy settings.

For instance, on Facebook, you can become a fan of Sandra Brown. Most people who visit her fan page really are not expecting to have a dialogue with Ms. Brown the way they would with their pals from college. This is a different kind of page with a different purpose. A Facebook fan page is a highly useful way for Sandra's PR people to post upcoming releases, new books, pre-order information, events, book signings, etc. on a social media platform that is directly integrated with Sandra's fan base. Not only will all of Sandra's 27,000 Facebook Fans be notified every time something new is posted, but those fans can then choose to repost that information, thereby extending Sandra Brown's reach beyond the boundaries of her fan page. This is much more effective than a static web site.

Additionally, someone doesn't have to manually approve every person who wishes to be a fan of Sandra Brown. This page also has the ability to give valuable statistics that will help Sandra's marketing people improve/tweak their marketing approach.

For published authors, the fan page is a far better choice and can be easily outsourced. But, even if you maintain it yourself, you will find that it makes interacting with fans far easier. Everyone will be approved, so you don't have to log in and do it by hand. There is a lot of streamlining afforded by a fan page, leaving you more time to keep writing brilliant books.

For unpublished authors, the fan page is a great step in a professional direction. I know what it is like for people not to believe you are serious. A fan page can help. Also, your fan page, if you keep regular content, will elevate your search engine ranking and help you brand your name far quicker.

All right, down to the nuts and bolts...

## Creating a Facebook Profile Page

### Step One—Sign up for a Facebook profile page (if you haven't already).

Facebook will need to verify you are a human, so when you sign up, you will need to wait on a confirmation e-mail before Facebook will allow you to progress. This is just one of the many things Facebook does to ensure that you aren't hassled by spammers.

### Step Two—Fill out all relevant fields.

As I mentioned earlier, you do have the option of including some very private information like your home address or personal cell phone number. I personally do not feel comfortable giving out that kind of information even if there are different security settings I can choose from.

******Facebook allows you to select who can view certain information. By clicking the symbol that looks like a padlock, you can choose the visibility—from visible only to you to visible to everyone on the planet.***

When asked to choose your permanent Facebook domain, I recommend using your real name. I'll discuss this a bit more later. ***There will be a section in your Contact Information where I recommend putting the links to your Wordpress blog,***

***all of your social media pages, and your web page if/when you have one.***

***FACEBOOK CONTACT INFORMATION WINDOW IMAGE***

## Step Three—Getting a Lay of the Land

Facebook has a lot to offer and entire books have been written about how to use it. I know, I think I read most of them. There are all kinds of bells and whistles and gadgets, but I believe a lot of that is more distraction than necessity. Most of FB is pretty self-explanatory and once you get set up I recommend pushing buttons and experimenting (not like you are going to accidentally delete the Internet). For our purposes here, we will just touch on the fundamentals to get you up and ready.

**Essentially your Facebook page has 3 main modes—Home, Profile, and Account.**

The **Home** page is basically the big picture. Home represents everything that is going on with everyone in your Facebook network (once you have one). If something is posted—a picture, a video clip, a status update—you will see it stream by in your **News Feed**.

***FACEBOOK NEWS FEED IMAGE***

Your News Feed will also allow you to see anyone who has commented on anything posted. So, if your pal from high school posts a hilarious video from You Tube, you can join in the conversation with others. If you hit it off, take a peek at their page and add them as a friend. These Facebook discussions are a great way to extend your network and get people knowing about you and eventually what you write.

The Home page is very helpful for keeping you organized and is dedicated to helping you grow your Facebook network. Facebook wants you to know as many people as possible and will even make suggestions for friends.

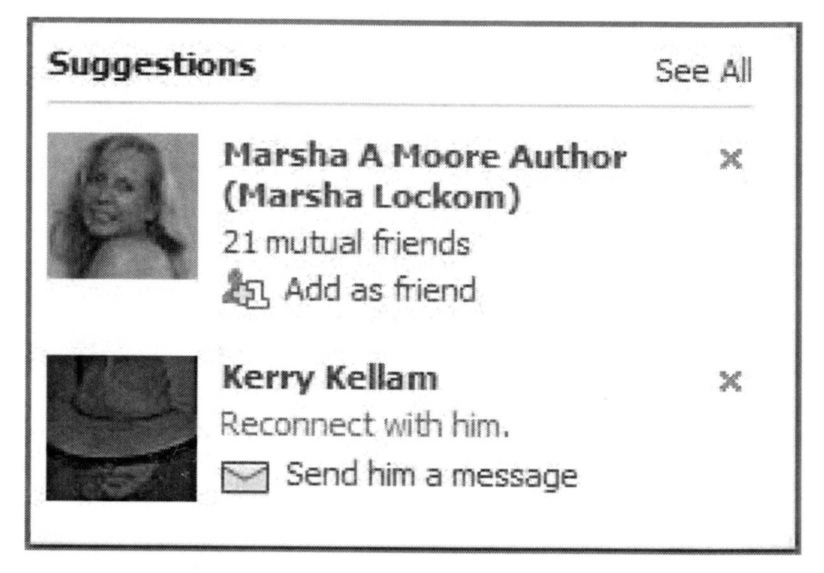


***FACEBOOK FRIEND SUGGESTIONS IMAGE***

Facebook uses the Home page to remind you of any unread messages, notify you if someone has sent a friend request, or even give you important reminders like upcoming birthdays.

The **Profile** page is broken into four main sub-tabs—**Wall, Info, Photos, Boxes**.

Your **Wall** is a compilation of everything that you have been up to in the wide world of Facebook. It will show your most recent status updates as well as any other recent activities.

RECENT ACTIVITY

Kristen likes Sandra Brown. Remove

Kristen and Gunter Kaesdorf are now friends. · Comment · Like Remove

Kristen and Nancy Pistorius are now friends. · Comment · Like Remove

***FACEBOOK WALL/RECENT ACTIVITY IMAGE***

Your Wall is everything on Facebook about YOU. Who you have befriended, causes you have joined, your status updates, messages to you. All that stuff about YOU is funneled to this page. When other people go to your Facebook page, it is your Wall that they see and it is the place where they will write public messages to you, much like a bulletin board. Your Wall gives them an idea of who you are and if it makes sense to add you to their network.

This also works the other way. You can go look at other people's Walls and use it to get to know them a bit more, establish a dialogue or even see about befriending their friends as well. Birds of a feather flock together, right? If you write vampire books and love YA and Twilight, then you probably would have a lot in common with friends of your friends. Capitalize on that.

You have the option of removing anything you want from your News Feed or your Wall by simply clicking on the word **Remove** off to the right. We all misspell stuff or post something then change our minds. No big deal. Just remove it.

Maybe you don't want your entire network to know some piece of information. There is a way to control what you portray to the public. Just click on the icon that looks like a magnifying glass to select who sees what tidbit of information. Only friends? Everyone? It's your choice.

The next two tabs really have to do with what content you are adding to Facebook.

The **Info** tab is for (you got it) information about you. This is where that bio I had you write earlier will come in real handy. Now all you have to do is cut and paste. As far as all the other boxes, feel free to fill in only what makes you feel comfortable. If you don't desire to share your religious or political affinity, you don't have to. I don't think it is a problem for you to admit you are a Democrat or a Republican, I just advise against posting content that is divisive. Ranting about the government is best left to others.

You can leave some sections for later. Letting others know your favorite books and movies is more for conversation starters than anything else. They help paint the picture of you to outsiders. Some of this information also helps people know who you are before they send a friend request. If someone searches for *Kristen Lamb* and twenty *Kristen Lambs* pop up, then whoever is searching can then use my information to narrow the field by looking for a *Kristen Lamb who went to Texas Christian University*.

**Networks** are also a way Facebook helps you connect with friends. Past schools attended, earlier jobs, etc. give FB more ways to connect you with people you might know. If you don't fill in these fields, you limit what FB can do for you. But no need to feel obliged and you can always come back and fill in these boxes later. The minimum you need is the basic info, your bio, and links to your Wordpress blog, your MySpace, and your web site (if you have one). Eventually you will also add the URL of your Twitter as well.

**Photos** is obviously the place where you will upload photos for the world to see. Yes, you can change the privacy settings on the photo albums as well, but if you are posting pictures of your kids or other private images that you don't feel comfortable exposing to the world, I recommend another, personal page. Or just make sure you are super careful to always set the privacy.

***Note*** FB does frown on multiple pages, but that is mainly to discourage spammers and bots. FB wants real people behind those profiles because it is interacting and posting content that makes FB fun. Keeping up with more than one profile is tough work. If everyone out there had five profiles they never used, then that just detracts from the FB experience. But, so long as you are a good girl or boy and don't abuse the privilege, you should be fine.

Back to photos. I do recommend uploading pictures. They help start conversations and also help people get to know you.

To create a photo album:

1) Go to your **Photo** tab.
2) Select **Create Album.**

3) Fill out the information—Name Your album, Location, Description (use tags generated earlier if appropriate).
4) Set the Privacy Setting.
5) Select Upload Photos.

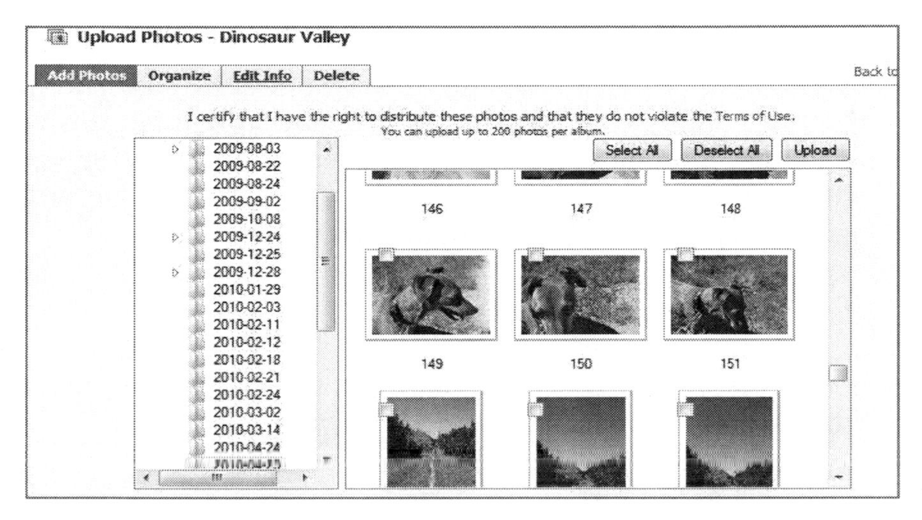


***FACEBOOK PHOTO UPLOADER IMAGE***

6) Select the file from your computer containing the photos you wish to upload
7) Hit **Select All** to easily select all photos in a folder, or check the small box in the upper left-hand corner of each image to select photos individually.
8) Click **Upload.**
9) You should get a pop-up box telling you if the photos uploaded successfully.
10) Facebook will then divert you to a place where you will see all of the photos you just uploaded. This is where you can add captions and also choose which image to make as the album cover.
11) Once you are finished, select **Publish.**

Feel free to upload pictures of you, picture of the last writing conference you attended, pictures of the place you scoped out to use as a setting in your novel, anything that represents you and your brand. You can also post pictures of your dog or your new rose bushes. Remember that personal side of you is important as well. You will have to just be mindful to keep everything balanced.

The final tab you need to know about is the **Account** tab. This is the tab that allows you to edit your settings, change your privacy, get technical help, and most importantly LOG OUT. If there are multiple people sharing the same computer, it is wise to log out of your social media sites or others can play havoc with your page (Just ask my brother how he ended up with a unicorn MySpace background, muah ha ha ha).

### Step Four—Making Friends

Once you have everything built and optimized, you will want to make friends. You can do this a number of ways.

### Friend Finder:

1) Click on your Home page
2) Off to the left side you will see Friends. Select Friends.
3) FB will offer a variety of options to connect you with people you know. The easiest is to search by e-mail

***FACEBOOK FRIEND FINDER IMAGE***

This feature will import your address book and let you know who in your directory is already a member of FB so you can send a friend request and get connected. Additionally this feature will give you an option of sending an e-mail to those in your directory who aren't already on FB encouraging them to join. What better way to start building a platform than by recruiting people who already know you?

**Poaching:** Birds of a feather flock together so go do some poaching. Go to your friends' pages and look who they have as friends. If they look like someone who would make a nice fit, then send a friend request. FB will tell this person you know someone in common, so they are much more likely to reciprocate.

You can also go to fan pages of authors who write a lot like you do. This is a bit riskier in that you don't have the person in common, but most of the time people will reciprocate if you are polite and don't come across as a psycho or a spammer.

**Groups:** Remember earlier we discussed fishing where the fish are. FB groups are great places to network, find friends, and create relationships with people who like the same things you do (potential readers).

You will have to do some work to find a good fit. Go to the search bar at the top of your page and type in what kind of group you would like to join. Say you are writing a book about a serial killer. "Forensic Groups" is a good place to start.

FB will then show a page that brings up all things **forensic**. Off to the left, you will see that there is a new list of options. The fourth option down is **Groups**. Select Groups and you will see that the list narrows to just Groups with Forensics in the title. Above the list, you will see a way to narrow the list even more. There is an icon that allows you to adjust which groups are shown—All Group Types. Narrow the field, and a new box will appear All Sub-Types.

Some groups are by invitation only. Some will make you send a request and wait for approval. Others allow you to join instantly and begin chatting away about posted topics. In my experience, a lot of the really large groups allow you to get started either immediately or pretty close to it. Once you are in the group, it won't take long to figure out if it is worth your time.

But, if you write about serial killers and you join a Forensic Criminology group, you are pretty likely to find a lot of fans who like books and movies about serial killers. Additionally, there is likely to be a lot of people who work or study forensics in this group. What a resource to have!

Meeting people on FB is very simple and FB goes out of its way to improve the process. Spend time building this profile while you are writing your book and then, once you get the book deal or a significant number of followers, go publish your fan page. Now you will already have an established network, so having an almost instant fan following will be far easier (and way less depressing) than starting from scratch.

### Step Five—Get Started

Go to your profile page and start interacting...after you go learn about Twitter. Once you get a Twitter profile and have everything linked together, you will eventually come back here and start interacting on FB. There is a long window at the top of your page that poses the question, "What's on your mind?" This is where you can share your hopes, thoughts and dreams, links to your favorite videos on You Tube or a link to you latest blog post.

To reiterate. I do believe it is important to have a presence on all four platforms. You will likely blog on Wordpress (or Blogger or your home web page) and then do most of your interacting on one of the main three—MySpace, FB, Twitter. You will need to spend time with each to see which you like the best (except YA authors—you guys NEED a MySpace presence whether you like it or not, so suck it up—*hugs*). Once you find a favorite, you can just go to the other two once or twice a day to check messages, comment on a status, or do a status update.

I recommend you ask yourself what tasks you can delegate, and who can you delegate to? Don't automatically assume **you** must do it all. You are a business. Do what businesses do. Outsource.

As I suggested earlier, pay your teenager who is goofing off on Facebook anyway an hourly wage to add quality friends to your page. Have your blog pulled up in a Word document and get them to post it and link it and send out the bulletin while you're cooking dinner. Now what you pay them is a tax deduction because it is outsourced work. Once you are published, you will need all the deductions you can get. Besides, don't you think your family would benefit from you being a famous, successful author? Recruit them to your team and allow them to not only support your success, but be a part of it.

There are also firms you can hire to outsource this work. For a monthly fee they will build your network. These firms will interact on your behalf with your content. There is no real substitute for special, unique you, but there likely will come a time that you just cannot do a good job maintaining or even expanding these networks on your own. Sandra Brown has to have time to write books, so she outsources. We will discuss this more at the end.

Now to build your fan page! Squeeeee! I love the smell of success.

## Building a Face Book Fan Page

You can easily build a basic fan page on your own.

1) Go to your web browser and type in **Facebook**.
2) Log in to your Facebook profile.
3) Go to your home page, the one with News Feed. Off to the left you will see your picture. Below your picture you will see **Ads and Fan Pages**. Click that.
4) Click the **Create a page.**
5) This will bring you to a list of choices. Select **Artist, Band, Public Figure** and you will see a drop-down menu will appear.
6) Select **Writer.**
7) Type in the name you desire to brand.

8) Now you will see that you have created a very basic FB fan page.
9) FB has now created a "Like" button instead of "Become a Fan" to take the pressure off of others. I guess becoming a fan was far too committal.
10) You will see that you only have 2 tabs—**Wall** and **Info** and there is a giant **?** mark.
11) Place your cursor over the ? mark, and an icon will appear that says **Change Picture.**
12) Upload your profile picture.
13) Below the profile picture, you will see a box that says, *Write something about your name.* This is where you insert your super-duper short bio.
14) Go to **Edit Page** so I can give you a preview and so we can change one thing. You will see this page has ways that you can control what age group is allowed to see your page (if you have adult content), who can post on your page, etc. I recommend looking through these and setting these to your preferences.
15) Under **Settings**, you will see a box for **Gender**. Well, Facebook uses pronouns and it helps to know if we are dealing with a *he* a *she* or an *it*. So set the appropriate gender. Most everything else, you can futz with later and it is generally self-explanatory. Just always remember that the fan page doesn't have the ability to restrict privacy like your profile page (another benefit to having both). Thus, what you post is for the world to see just like a web site.
16) You will also see above the gender box a **Published** box. Select **Unpublished (visible to no one but admins)** so you can build in peace.
17) Click **View Page** and holy, moly! Two more tabs have appeared—**Photos** and **Discussions**. You upload photos just like you did when you were uploading to the profile page, so this will all be familiar. Post your gathered content photos we discussed earlier in the book. You can use pictures of your book, book

signings, pictures of you at a conference or a writing group. If you are new and unpublished, get creative. Upload pictures of you speaking if you are NF. For fiction? Upload images of the place where your book will be set or artwork that portrays the world you are writing about, pictures of what inspires you.

**Discussions** is where things can get fun. Remember all that lecturing I did earlier about blogging about writing and finding an agent and getting writers block? HERE is the place for it. Post these on your discussion board. Hit Post Topic, and feel free to talk about coffee and conferences and bookmarks and a movie you found inspiring. THIS is the place where fans come to know about YOU THE AUTHOR.

Then, when you get ready to post your **blog**, post only the title and the first few sentences to whet the reader's appetite, and then the link to your **official blog** where you **blog on topic.** Now your fan page is well, for the fans. They can connect with you and have a window into your experience as a writer, but not at the expense of you connecting via your topic.

18) So if you already have one of your Wordpress blogs posted, go ahead and post it as a topic. Just enough to get people interested and then paste in the link. I advise getting a handful of these before going live. Remember, you want to appear like you are an old hat at this. I won't tell.
19) Click your **Info** tab. Click the pencil icon that says **Edit Information**. Fill out the relevant fields. Under **Basic** info, I would ignore most of it. Other people don't need to know your address, unless you happen to have a nice P.O. box. **Never, ever fill out birthday information! Ever!** If people are your friends, they will know your birthday or they can e-mail and ask. Identity thieves love getting their grubby little too-good-to-get-a-real-job-like-the-rest-of-us fingers on birth dates.

20) Fill out the **Detailed Information Box** with the content we have already prepared (using your tags). In the **website box** you can put the web address of your web site (if you have one), your Wordpress URL, your link to your MySpace page, and eventually your Twitter. This is so people can connect and interact in THEIR comfort zone.

I am on Facebook, but I LOVE interacting on Twitter. Remember fan pages are unidirectional, so I can't chit-chat with you on a fan page like I can Twitter. Help me FIND you so I can connect.

21) Under **Contact Info** just use the e-mail you have selected for business.
22) Select **Save Changes.**
23) You will see there is a tab with the + sign. This is where you can add your favorite links, video, etc. If you have any relevant video that counts as content, add it here. If you are a NF author and there are videos of you speaking? Load 'em up!
24) When you are ready to go live, click **Edit Profile.** Go to **Settings.** Remember we set it to a private view earlier? Just select **Published** in order to make your page visible to the world.
25) When you are finished, with everything else, it will be time to invite friends. You will be so grateful I made you wait until the Golden 300. Select **View Page**. On the left side below your profile picture you will see **Suggest to Friends.** Select this and all your 300+ friends will appear. Now you can invite them ALL! Wheeee!!!! And now you will have more than your mother and a third cousin as your fans.

...I love this job.

But right now you aren't ready for that. Flesh, flesh, flesh. Before you go live, you need to back through and fill in as much as you can. What would you do if friends were coming over to your new house? You'd buy furniture and hang paintings on the walls, maybe even put out vases of cut flowers and stock the fridge (or the bar depending on your friends). You would spruce the place up so as to impress your visitors. Your fan page is no different. In all fairness, what's the rush? Wouldn't you rather make a good impression? The Golden 300 is just a good number to shoot for, but when you decide to publish your page is up to you.

For now? Don't publish until we finish up with Twitter. Stay focused. You might notice that when you go back to your main page, that FB offers to link with your Twitter. Yes! Yes! Yes! Make life easy. Problem is? You need a Twitter account first, so let's get to work!

## Twitter

Ah, Twitter how I love thee? Let me count the ways.

Yes, I love Twitter. I am addicted to it and couldn't live without it. Twitter, in my opinion, is the perfect social media platform for writers. At first, I guarantee that all of you will think that I am crazy. If you don't know the inside skinny on how to use Twitter, it can be difficult to find the experience appealing at all.

Ah, but once properly indoctrinated, you will be our slave...I mean Twitter friend. Yeah.

One of the big reasons I love Twitter is I hate reading instructions. Honest. Which is actually kind of ironic in that I made a living for many years writing—irony of ironies—instructions. I seriously have a phone that could raise my child for me if I downloaded the right app. Frankly, I am not that motivated, and in some things am actually quite lazy. Okay, maybe I'm not lazy, but like you I have a writing to do, a blog to update, a family to tend, a house to clean, a dog to walk, a cat to feed, a car that needs an oil change, a yard that needs weeding, and...

You get the point.

To me, Twitter is what ties all of this magnificent social media work together, and is a powerful tool for dominating the domains with the minimal amount of effort. It is truly amazing what you can do with Twitter if you know how to use it properly.

So go to Twitter.com and sign up for a Twitter account if you haven't already. Make sure you use your brand or some variation thereof. Twitter will ask you for a username and you will, of course, use your name. Twitter will check for availability. Try to get as close as possible to your brand name. If you have to, sacrifice your first name. When we go to buy your book one day, your book will be shelved by your last name, so that is most vital.

Twitter will make you agree to the terms of use and then will send you an e-mail to confirm that you are a real live flesh and blood human being. Then once you click the link to confirm that you are in fact human, you will be directed to your brand spanking new Twitter page.

You will be tempted to tweet. Don't. You will be tempted to follow Twitter's coaching, and find friends. It's tough being alone. I know. Twitter will, like Facebook, be super accommodating at helping you import your e-mail address book and locate all your pals on Twitter and invite others who have yet seen the Twitter light. Don't. You have work to do first.

ShawnCMac
0 tweets
0 following 0 followers 0 listed
blip·fm
n. a easy way to share music on Twitter.
Home
@ShawnCMac
Direct Messages 0
Favorites
Retweets
Search

***BLANK TWITTER PROFILE IMAGE***

If you start tweeting, who is going to see what you have to say? You have no followers. And if you start interacting with other people, you have no profile picture and no bio information. That's like talking to strangers...creepy strangers. So let's remedy that.

**Twitter Instructions**

1) Up in the upper right hand corner, you will see a Twitter toolbar.
2) Select Profile.

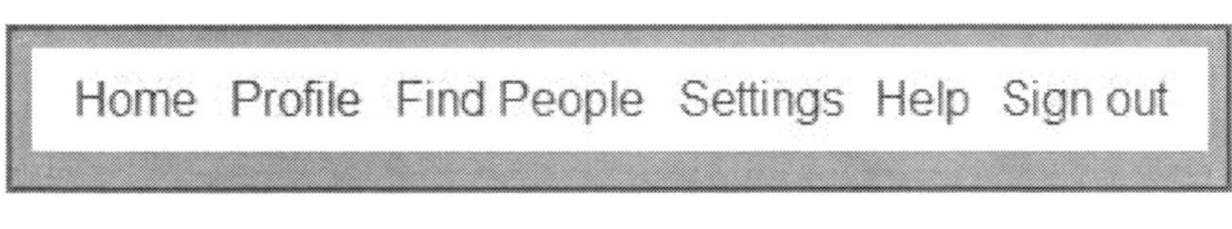


***TWITTER TOOLBAR IMAGE***

3) Click the box next to your name where a photo will go.
4) Upload your image.
5) Put in your location.

6) You will see a place to put in your web address. You can choose here which site you would like to display. Your MySpace, your Wordpress blog, or your static web site. I would choose MySpace if you don't yet have a fancy optimized site. Once you have an optimized web site that will be the wisest choice (an optimized site will have widgets that will allow others to connect with you on their social media platform of choice).
7) You will now fill out your bio. As you will quickly see, you can only use 160 characters. That was why I wanted you to write a super-duper short bio. See how much time you saved because it was done ahead of time?
8) Click **Save.**

9) Look to the top of the page and you will see a new menu has appeared. You should be on **Profile**. Next to **Profile** you will see **Notices**.
10) Click **Notices** and Twitter will bring you to a page where you can decide whether or not you want to be e-mailed when someone Follows you or sends you a direct message. I leave it checked, but that is likely why I have 1600 unread messages. We will talk more about Following, etc. later.
11) Hit **Save**.
12) Now click the **Design** button. We are going to jazz up your Twitter page. If you didn't find anything earlier that you like, Twitter has some basic backgrounds to at least make your page a little more unique than the standard blue sky.

Or, go back to the page that you bookmarked on Tweetygotback.com. Tweetygotback makes changing your background easy squeezy. Log in (it will have you give your Twitter username and password). Find the background you like (you might have already found this and noted where it was).

Tweetygotback will allow you to Preview how it will look. If you like it, click **Apply**. That's all!

Once you get the hand of all this social media stuff, I do recommend paying for a customized background that will allow you to use more of your Twitter real estate. On www.twitbacks.com, you can design a Twitter background that will allow you to use that left-hand open space.

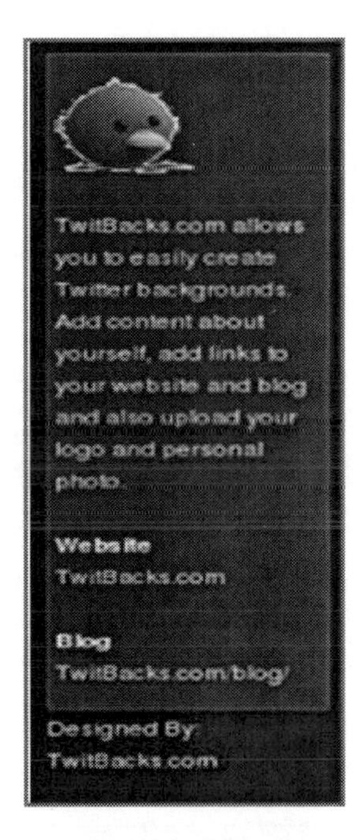


***TWITBACKS IMAGE***

As you can see, Twitbacks allows you to use that valuable Twitter space to give links to your blog, your web page and even a longer bio. You can also upload a photo and a logo. Like these other sites I have shown you, they have made it super simple. Just follow the cues and click. Twitbacks you pay for, though. Most of the time, it isn't that much, a couple of dollars. Tweeygotback is free and has MySpace backgrounds to match.

13) You may have a spiffy page, but this is a page for the casual tweeter. You will eventually be keeping up

with THOUSANDS of people, so this singular column is just inefficient. You are about to download the single greatest tool ever bestowed upon writer-kind...The TweetDeck (cue dramatic music).

14) Open another browser tab.
15) Go to http://www.tweetdeck.com/.
16) In the upper right-hand corner you will see **Register.**
17) Follow the cues and download TweetDeck.

**TweetDeck Instructions**

You are going to LOOOOVE TweetDeck. Well, at least I hope you do. Once you log in and open Tweet Deck, you will see that this large black screen of columns has now appeared. This is how you are eventually going to manage who you are following as well as find new people to follow and to even connect with potential readers. This application will allow you to follow thousands of people and actually be able to keep up with them and interact. Seriously.

TweetDeck, like other applications, changes quite often, so I am only going to go over the basics. Guaranteed there will be some new doo-dad before this book is even finished. So we are going to keep things simple.

On the upper left-hand side, you should see something similar to this:

***TWEETDECK MENU BAR IMAGE***

The yellow button is what you press when you want to write a tweet, which is basically a status update (140 characters or less). Next to that yellow button is another button with an icon of a magnifying glass. This icon is your new best friend, and we will discuss why here in a moment.

With a brand spanking new TweetDeck, you will see that you only have a couple of columns.

**All Friends**—everyone in the Twitterverse you happen to be following. I let this column stream by and scan for anything that catches my eye.

**Mentions**—is any time someone on Twitter mentions your name. This is handy to help you see when people repost your tweets. It alerts you to thank them or follow them in kind or whatever.

**Direct Messages**—are messages to you, personally. You have to be following the person who sends the message. In turn, if you want to send someone a direct message, they need to be following you. No one but the two parties involved in the message exchange can see the content of the message. So if you need to say something to someone that you don't want the whole world to see, this is the place. I have given people my personal e-mail or my phone number and that is something I would rather not post to the world.

**Favorites**—These are the tweets that just warm your heart enough to want to save them. I saved the times that Writer's Digest Magazine named my blog as one of the best. I have also saved links to things I thought were funny or interesting. This is the place to save the tweets that, for whatever reason, you want to keep.

Beyond these four columns, it is up to you to organize everything, and TweetDeck makes this simple by allowing you to organize people into groups. Once you have some friends (we will get there in a minute) you will see they likely will pop up in your **All Friends** column.

***TWEET IMAGE***

This is my pal, Fred. You see his profile picture, his tweet, and some writing at the bottom. If you move your cursor over the Fred24Live and click, it will take you to Fred's profile.

***TWITTER USER PROFILE IMAGE***

This is how you get to see if the person is someone you want to follow. Do they have the same interests? Do they fit into the profile for a potential reader? Are they a fellow writer? We will talk more about this in a bit.

If you look at Fred's profile, you will see a lot of information about Fred, namely his name, Fred Campos, his bio, where he lives, how many people he is following vs. how many are following him. That is an important number to pay attention to find your *referent influencers.* If this person tweets a lot and they have over 2,000 followers, then you want to take good care of them.

Fred also is kind enough to warn you that he tweets a lot, so you know if you follow him that he will pop up quite a bit—hallmarks of a *referent influencer.* Fred has also listed his website.

When people click on your name, all that bio information I had you prepare is what they will see, just like here with Fred.

What about the fancy background?

Not everyone is on TweetDeck, so if they view your profile just normally they will see the graphics you took time to upload.

Back to TweetDeck...

Once you do some tweeting, you will go back to the Twitter home page and download your e-mail contacts. For now we are going to go fishing.

## Separation—Creating Search Columns

Remember that button with the magnifying glass icon? Click that and a bubble will appear.

***TWITTER SEARCH COLUMN IMAGE***

When starting out, you might want to just gravitate to birds like you, so go ahead and type in Writer, Publishing, or something familiar. You should see a new column appear chock full of twitterers. What the Search column does is it filters all 65 million tweets per day and any time that word you selected pops up, TweetDeck puts it in a column.

Do a quick scan. Do you see anyone on there who sounds interesting? Okay, well click their name below their picture and look at the bio. If this person looks like someone you might like to follow, hit **Follow**.

Now you will likely see this person is in both columns. They will appear in the **Search** column (i.e. Writer) AND they will pop up in your **All Friends** column. This person likely will be notified you are following them, and most of the time people reciprocate.

I love Search columns and I have them for all sorts of subjects—writers, screenwriting, Rotary International (I am Rotarian and I use TweetDeck to help make friends with Rotarians around the world), agent, publishing, etc. scan the column and see if the search word is catching good fish, then I pluck out the ones I think are interesting (I follow them). Sometimes search words don't find the kind of fish you desire. For instance, I started with **rotary** as a search word, and Twitter delivered everyone tweeting about rotary phones, rotary engines, so I had to try again.

Remember earlier I had you profile your reader? This is where your search columns will help you befriend people interested in your topic. If you are writing a book about space aliens, create columns for UFO, Star Wars, Star Trek, pyramids, Easter Island. Profile your potential reader. What would he or she be likely tweeting about other than your book? Befriend writers, but unless you are selling books about writing, it is wise to expand your circle. Use the profile I had you create earlier to create your Search columns and Follow away! Become friends with people interested in your topic.

At this point in the game, it is a good idea to watch and get a feel for things. If someone posts a tweet, and you feel you simply MUST respond, go ahead. Place your cursor over the person's picture and you will see that the profile picture will turn gray and divide into four quadrants. The upper left quadrant has a curved arrow. When you click that arrow, that person's name will appear in your message bar at the top of the page.

If you replied to me you would see:

***TWITTER REPLY IMAGE***

Okay, so I typed the message for you. You're welcome.

You can also just type the person's name with the @ symbol in front. The great part about Twitter is that you can talk to anyone you want so long as they are on Twitter (duh). I have replied to posts by Kim Kardashian, Chelsea Lately, James Rollins and even Wes Craven. Your comment will appear in their main column (if they are on regular Twitter) or will appear in **Mentions** for an application like TweetDeck. Sometimes, they even reply! Twitter lets you follow whoever is on Twitter with an open profile (some people protect their tweets, which is just odd and not something you should do). Most people have an open profile, so you can follow them

Find mentors! Search for agents and publishing houses and editors. Search for famous authors and learn from them. If these people take the time to post a link or a blog, that is likely something valuable. We discussed earlier how a tweet by NY Times Best-Selling Author Susan Wiggs led me to my favorite book to teach structure. Follow the best. Follow winners. Follow the 5%ers and learn what they do and how they think. You have to study success to be successful. You become who you hang around. Use Twitter to learn what they value and how they structure their day and if that is something that would work for you. Some big authors write 5 pages a day (Elizabeth George). Others write 3 hours a day (Jodi Thomas). Learn as much as you can and apply what works best.

Search for all the best-selling authors you love. Search for the ones who write books similar to yours…then follow their friends. If you write like Sandra Brown, the people following her might eventually like your book too. Who are they talking to? Click on their profiles and follow them too.

Fish where the fish are.

These **Search** columns are how you can follow and join in those hashtag conversations. Remember earlier I said that these were Twitter's answer to groups? All you have to do is create a search column for #writegoal to have an entire column dedicated to following this discussion.

Some popular writer hashtag discussions are #writegoal, #amwriting, #nanowrimo (National Novel Writing Month), #pubtip. Every Friday, people on Twitter engage in what we call Follow Friday. This is where people on Twitter recommend other people. This helps narrow down your search of whose tweets are worthwhile. If @Bob_Mayer, @jamesrollins or @christinadodd bother to Follow Friday people, you bet I am following them. People in my network whom I know and trust are literally providing me with a list of referrals. If they are taking the time to do this #FF, it is wise to listen.

***TWITTER FOLLOW FRIDAY IMAGE***

These hashtag conversations are also great places to connect with others. I teach writers, so I follow all of these hashtag conversations listed above. If I see someone tweeted that they made their 1000 word goal, I reply, "Congratulations" and I click their profile to see if I am following them. If not, I click **Follow**. All this person knows is that someone just took time to congratulate them. Very likely this person will click on my profile and follow me. And so on and so forth.

If I want to participate in a hashtag conversation, I type my message and then a hashtag at the end.

So I would type: My goal is 1000 words before lunch #writegoal

My message would appear in the #writegoal column and all the people all over the world who are following this column would see that tweet appear. These hashtag conversations have helped me make friends in the UK, the Netherlands and even Australia.

If somebody posts an interesting tweet, something funny, informative, etc. then feel free to retweet them. Move your cursor over their profile picture and you will see those four quadrants. The left side bottom square is a →. Hit that and **RT@*Their Name*** will appear in your message bar above. You might have to delete some extra words (the bar will turn red if you are over the 140 character limit), but once you are ready, hit **Enter.**

**RT**ing is how you can extend your reach exponentially. When I post my blog, it goes out to my followers. But if Bob Mayer RTs, it goes to his thousands, and then others RT and you can see how one piece of information could quickly travel the globe.

If you are engaging in conversation, people generally will be kind and follow you in return. It won't take long before you can have quite an impressive entourage. TweetDeck makes it possible to connect with almost anyone and keep them separated and organized.

Remember earlier we discussed the perfect tweeting proportions—1/3 Information, 1/3 Reciprocation, and 1/3 Conversation.

I leave TweetDeck minimized when I work. When I take a break, I scan the columns for anything I could RT, then for someone I can congratulate or interact with, and then I post my blog or another piece of valuable information. Sometimes I am more heavy on the conversation and I have to remember to back off and tend to those other parts of the equation. I am mindful to support my friends and acquaintances and repost their blogs for them. This is how relationships are made.

## Organization—Creating Groups

Now that you have Search columns, you will likely also want to organize your people into groups. If you have writer pals that you love keeping up with, then you need to create a special column for just them. I have a column on the far left-hand side labeled **Close Friends**. If you make it into this column, it is because you are a close friend, or I find your tweets interesting enough that I don't want to miss them. This is the column where my favorite best-selling authors are stored. This TweetDeck function allows you to keep up with family, friends, politics, business, church, writing, current disasters or celebrity mishaps...all on the same screen.

### How do I create a group?

In order to create a group, go back to your magnifying glass icon and click it. The bubble will appear (this is where you wrote in your search word). You will see a list of icons at the top of this bubble. Roughly the fourth one in looks like stacked human silhouettes. If you slide your cursor over that you will see "**Groups**" appears. Click that and this appears:

***TWITTER "CREATE A GROUP" IMAGE***

If you are already following people, Twitter will put all of them in a column in alphabetical order. Check the box next to people you want put into this group and name it "Writers," then hit **Save Group** (gray button located at the bottom). Now Twitter will put the people you selected into this group.

What if someone in one of your Search columns is just really interesting and you want to move her into your group of **Writers**? Place your cursor this new friend's profile picture. When it turns gray, you will see an icon on the lower right-hand side that looks like a **cog**. Move your cursor over that and a pop-out menu will appear and give you two options—**User and Tweet.**

***TWITTER "ADD A GROUP" IMAGE***

Follow the prompts and add her to your group noted **Writers.** Now your new writer pal will pop up in your special group column. You will also see her pop up in your **All Friends** column, **and** in the original search column where you found her if she happens to use the term you are searching for.

So I have friends who will appear in **Close Friends, All Friends,** and **#writegoal.** Not that this is anything to concern you, but I didn't want you to panic and think you had done something wrong. If you stop and think about it this makes sense. **All Friends** is everyone you are following. **Writers** is the special columns you have slotted this person into, and **#writegoal** is obviously used by this person in that this is how you found her to begin with!

There is no need to feel overwhelmed. Most of the time, you will only pay close attention to one or two columns. The others you will likely scan and just see if something jumps out. That is part of the reason that you need to make sure your blog titles are catchy, and if they aren't, then the tweet you use to lure your readers better be. This was why I took time earlier in the book to help you understand basic marketing and not to speak to people's emotions and curiosity. It will help you stand out in a sea of competition and be reader-centered.

I minimize TweetDeck. When I take a break from writing, I pull up my Deck and scan the columns for anyone or anything interesting. If a blog catches my eye or a link looks interesting, I will indulge and read. If it is any good, then I will RT for others to enjoy. If the person looks interesting, I will follow them. If I have friends who have posted their blogs, I will RT for them and help them out. I have writer friends who have books for sale, books being released, etc. They teach workshops they need to fill and have book signings to announce. When I see they need help spreading the word, I RT. This is how you eventually sell books and drive traffic to your out of town book signings.

The title of this book says it all. *We Are Not Alone*. Social media makes it to where you now have help. You don't have to do it all by yourself. And trust me, I have seen this system do miracles.

## Using Twitter to Drive Traffic

Once you get to a certain point, you will want to drive traffic to your website or your Wordpress blog. You can do this easily with Twitter. Type your message and then paste the link in the message bar. Twitter will automatically shorten the link so you have room left over for a message.

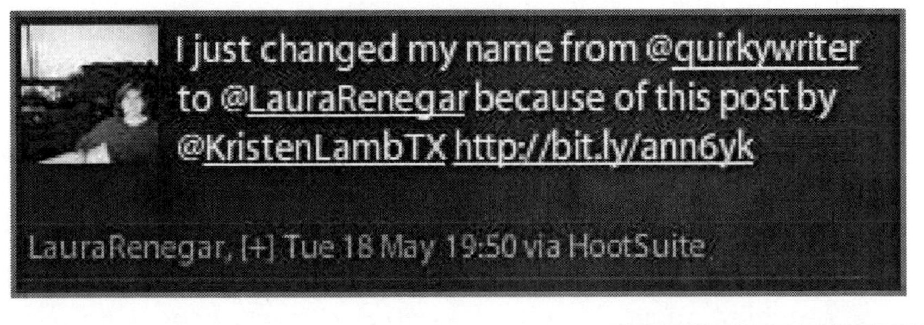


***TWITTER POSTED LINK IMAGE***

When you work as a group, there is no limit how far your message can go. This past year, the DFW Writers Workshop used this system and we sold out our conference two days after early registration and *months* before the conference. We not only sold out and sold out early, we sold out in a bad economy when other conferences were suffering.

Because of this system I helped put in place, the DFW Writers Workshop had massive exposure, exposure you can't BUY from a PR firm. We all worked together to promote. Didn't take a lot of time. Just hit the arrow and RT. Poof! Done. Write a message and paste the link to the conference web site. Takes two minutes. Poof! Exposure.

The beauty of Twitter is the power of the collective. Need a question answered? Tweet it. You will be shocked how many people will reply in under three minutes. We writers deal in information (even fiction writers). Last year I was working on a fiction novel. I was toying with the idea of making my female protagonist a bounty hunter, but I wanted it to be realistic. So I tweeted, *Can anyone tell me a good resource to learn about bounty hunters?* I actually had **bounty hunters** reply and offer information (they probably had a search column for **bounty hunters**). I had people send links and names of people. Talk about saving time.

Recently I had a computer issue, so I tweeted. *Help! Computer Savvy Tweeps. How do I blah blah blah?* Within thirty seconds I had a half a dozen replies. Some sent links. Some DM (Direct Message) me their phone numbers and offered to walk me through.

There are all kinds of stories of people using Twitter for similar purposes.

*I am in south Arlington, TX. Who wants to go eat?*

*Stranded at Reagan International Airport and need a ride. Help!*

*Do I bake or broil a turkey? My wife forgot to say which.*

*Found this kitten and she needs a home (attaches picture).*

*Teaching Social Media for Writers this weekend. One slot left (insert link).*

Twitter is a great place to poll opinions. Last year a friend of mine was starting a new fantasy novel and she needed feedback. She'd come up with a name for her bad guys and she posted it on Twitter for feedback. She called them *Necroids*. After about a dozen of us tweeted back and told her that it sounded like something in need of a healing cream, she decided to try something else. But what a great way to get instant feedback.

Twitter is extremely useful for driving traffic where you want it to go. Send people to your blog, your web site, your fan page. It is up to you. Use Twitter well and your name can spread across the globe in months. I have followers for my blog from Scotland, UK, Germany, Denmark, Italy, Switzerland, and even Kenya.

We have one more section and to go. This will link everything together so that no matter where your potential readers congregate, they can connect with you. Let's go optimize!

## Stage Three Optimization—Bringing It All Together

Optimization is what is going to thread all your hard work together. To make it really fancy, I recommend that you outsource. But, here we will at least go over the basics, and for those who feel more industrious, I recommend http://www.widgetbox.com/. It is a cool site that is fairly simple to follow. You can even make your own widgets. There is just so much you can do and so many variations that it would make this book too long.

Buttons are going to be your best friend, and they are so simple you could do it with your mouse tied behind your back. What is a button? Well, essentially it is a widget that will allow people to follow you wherever THEY feel most comfortable. So say a potential fan runs across your nifty blog about the pyramids and space aliens. He sees that there is a Facebook button embedded in your blog. Well, guess what? If he spends a lot of time on Facebook, now he can befriend you in HIS comfort zone. On my Wordpress Warrior Writer blog I have buttons for Twitter and Facebook.

So we are going to keep real basic and just make sure we install buttons on everything we can. A site with some really nifty buttons? http://www.socialmediabuttons.com/. This site is soooo simple. Just type in your Facebook username (what's in your FB URL) and then select the button you like. Hit copy, then paste...for now? Paste the code in a blank Word Document. I like to keep all my code stuff in a folder so if something happens, I don't have to try and remember where I found what. Copying code into a blank Word document will also make it where you can build your pages whenever you want. You can get the code now and then tomorrow it is in a place where it is easy to retrieve. Repeat the process with the Twitter button.

Twitter also has buttons that are really neat. Go to Twitter's home page www.twitter.com and scroll to the bottom of the page, where you will see **Goodies.** Click **Buttons** then select the one you like and copy the code in the same fashion. www.twittericon.com also has some nifty Twitter buttons.

Okay, distracted by the shiny things. Back to work. Once you have the code, go open up your Wordpress. Time to soup this baby up.

### Optimizing Your Wordpress Blog

1) Log in to your WordPress.
2) Go to Your Dashboard.
3) Select **Appearance** (located in the left-hand column, third block down). This is where you found your Wordpress blog background.
4) You will see a miniature version of the background you have chosen. Next to that, you will see Options, and next to that is **Widgets**. Select **Widgets.**
5) There will be all kinds of extra goodies you can add to your page in the **Available Widgets** area. Off to the side of that area, you will see a gray box that is now empty.
6) Move your cursor over the RSS. The RSS feed is important. It will allow your followers to just have your blog delivered to their news feed like pizza delivery. Click, drag, and drop the RSS into the

**Sidebar** section. Repeat this process with other extras you would like to have on your page. I would definitely add the widget that allows people to subscribe, archives, and most importantly, a way to connect with you on all the other pages. Remember that widget you found earlier for your Twitter and Facebook, the funny code?

7) Drag **Text** into your Sidebar. Click on **Text** and it will open a box. Copy your Twitter code out of the document where I had you paste it. Now hit **Save.**
8) Drag another text box and repeat with your Facebook.
9) Drag another text box and copy in your MySpace URL. I looked and looked for an easy button, but couldn't find one. I even consulted with my computer genius pal, Renee Groskreutz who built my web site, and she actually had to go write code for it. So, if you want an actual MySpace clickable button, you may need to outsource unless you are very computer savvy. Otherwise, just put in the link. Visitors can just highlight and paste in the browser.
10) If you decide you don't like any one of these widgets or want to table one or more of them for a while, there is a box below **Inactive Widgets**. Just click and drag the widget into this box and it will no longer appear on your blog page.
11) You can click and drag to move widgets up and down to suit what looks best. I recommend the tell-tale RSS box be left at the top, though.
12) Go check out your page and make sure all looks the way you want it. Now when people go to your blog, they can get it in an RSS feed, they can subscribe, and they can also follow you on Twitter, MySpace and Facebook
13) Blog on topics that allow you to insert hyperlinks. This will help broaden your network.
14) Approve trackbacks to your post. Allow people to reference your blog. Feel free to double check before

approving, but trackbacks are valuable for optimization.

## Optimizing MySpace

1) Log in to your MySpace and go to your dashboard.
2) Move your cursor over **Profile** and a drop-down menu will appear. Select **My Blog** and then off to the left-hand side, below your picture, you will see **Post New Blog.**
3) I recommend posting one or two of your shorter blogs from your strategic content in the blog section, just to give visitors a feel for your writing. The most recent entry will be at the top, so in that slot, I suggest you paste a copy of the link to your blog in the actual blogging section. Title it, "Link to the *Your Name* Blog." If visitors want to go right to your blog, they just click the link. If they are a little unsure, they can read your other blogs first.
4) Once you're finished with that, it is time to go install buttons. Go back to your dashboard (hit Home) and move your cursor over **Profile.** When the drop-down menu appears, select **Edit Profile.** Maybe you remember this place. I recommend going to the **Who I'd Like to Meet** box. Copy and paste your Twitter code, hit the space bar a couple of times to delineate the code, then copy and paste your Facebook button's code.
5) Hit **Save.**
6) Go take a look and make sure you like how it looks. You may try placing it in other boxes.
7) Make sure if you have a static website, that you put the address to your website either somewhere in the information (though it won't be clickable), or you can paste it in your blog section, then it will be clickable directly to your site.

Now if someone runs across your MySpace, but then in two months decide they use Twitter more often because it is easier with their new cell phone, you have made it simple to follow you easy-squeezy. They can interact with you from any of the major platforms, and easily find your blog.

8) I would also recommend, at this point, syncing to your MySpace to Twitter. You can set it up to update your Twitter every time you send out a MySpace status update, or you can do the opposite and update your MySpace every time you tweet (recommended). This is still no substitute for logging in and making a once-over. Every day I log-in, add a couple friends, acknowledge any requests, type in a status update, reply to any comments, and say happy birthday to any birthdays that are noted on my dashboard. Takes five minutes and it is invaluable for extending your social web. To sync your platforms, go to **Edit Profile,** select **Account Settings**, then select **Sync** and enter in the prompted information. Now when you tweet, it will update MySpace or vice versa.

### Optimizing Facebook

For both your profile and your fan page, just make sure all three other sites are listed in your Info section—the URL link to your static site if you have one, your Wordpress blog, your MySpace, and your Twitter. Yep, that easy.

### Optimize your Twitter

In your bio, either use your static web site link or your MySpace. I recommend spending a little money (a couple of dollars) and getting a Twitbacks background because then you have that extra space in the background to put all of your links, FB, MySpace, Wordpress (and a static site).

You can also sync your Twitter to update your Facebook. On your TweetDeck in the upper right-hand corner there is an icon that looks like a wrench (**Settings**). Click the wrench and select the **Accounts** tab and **Add New Account**. Add your

Facebook and MySpace, by following the prompts. Now you will see that above your message bubble where you tweet, some new boxes have appeared, that say Facebook, MySpace, etc. Highlight which ones you desire to tweet to. So if you just looove TweetDeck, which I KNOW you will, now you can interact with your MySpace and Facebook community as well, all from the same place. No more logging in and trying to juggle all of them.

Again, I still recommend that you log in daily for a once-over, but you can post all kinds of content from one spot...your favorite. And your fans and readers can interact with you from one spot....THEIR favorite.

Everyone wins.

With everything optimized...it should be relatively easy to build a solid platform. You will have to work and work hard, but your efforts will have more impact because now you are employing an efficient machine. This will be the difference between using a hand shovel to dig a hole and a back-hoe.

Well, now you have everything built. What now? Glad you asked...

## Act III—Managing Your Social Media Platform
## Time Management

Social media is like a lot of things in life. It is a collection of small actions that will amount to a huge difference. Employ small amounts of meaningful activity every day and you will be amazed how quickly this will grow. You have the help of others, remember. The 2010 DFW Writers Workshop Conference sold out not because of a handful of large, expensive, time-consuming actions. It sold out because countless people making countless small actions toward a common goal made a big impact.

You will have to invest your time in it for it to prosper. How much time is up to you, but with PDAs getting cheaper and cheaper, there seems little reason other than an illness or death can reasonably keep you from at least checking in daily. I never let more than two days go by that I am not on social media. If I am away longer than that, something *big* has happened.

If I know I am not going to be able to be on social media, most of the time, I will at least post something to ease the minds of those following me. *I have a massive deadline. Sorry guys. Can't be distracted.* If you are participating on social media regularly and in the way I suggested, you will be missed if you disappear. So try to be polite and at least call home. We worry about you.

How do I manage to accomplish all I do on social media?

### Maintenance/Communication

I control most everything from TweetDeck, but I always have Facebook and MySpace open. I minimize all of them. Throughout the day when I finish some work/writing task I will maximize Twitter first. I will scan the columns for all three types of tweets—Information, Communication, Reciprocation.

First, I will post a blog or article. I will look for something to reply to conversationally, even if it is a congratulations or a word of encouragement for some random writegoal# writer. Next, I actively look for people to cheer up, support, make laugh, etc. Finally, I will scan for people who posted their blogs or book-signings or new releases, and I will edify them and RT them to help extend their influence. I will also make sure I reply to any messages directed at me and will thank anyone who has reposted my information.

Easy. Takes five minutes.

Now I might not always go in that exact order, but I do make sure I try to keep a balance of all three types of communication.

Once finished on Twitter, I go to FB and repeat the process and then MySpace. Scan the News Feed/Stream for anything I can comment on. I also check for messages that might need my attention. Again, five minutes.

### Adding Friends

Generally I will just add friends as I go on all three. Sometimes, I just have to sit and carve out time to increase my numbers...or delegate it to someone else. On Twitter, I generally add all the FFs that pass by. Sometimes I will just follow anyone in a search column who looks cool. Easy.

On FB, the **Home** page will always offer you suggestions based on common friends. I just keep sending friend requests until Facebook runs out of suggestions. The more friends I have, the more suggestions, and the more people we have in common. I have FB suggest friends who have 206 friends in common. Now THAT is a network. Also see if you can add the friends of your friends.

On MySpace, I recommend going to other people's pages and poach their top friends, but I also suggest you go to an active MySpace group and befriend people who have recently posted on a topic. This way you are befriending people who actively use MySpace.

**Blogging**

When I post a new blog on my Wordpress blog, I announce it on Twitter, Facebook and Myspace. You can do this via the TweetDeck, or you can log in and physically post on all three. You will have to log in to post on your fan page anyway. When it comes to posting on a fan page, I suggest that you only give a tantalizing couple of sentences and then the link to your blog. Post on your regular FB profile and the fan page, even if you haven't published it yet. That will help you flesh it out little by little so you aren't killing yourself by trying to do it all at once.

If you have all three—MySpace, TweetDeck, and FB minimized, then it is simple and fast to just post on all of them—bam, bam, bam. And since you will be actively participating on all platforms, your friends won't feel like social media stepchildren.

In the end, you don't have to be as interactive as I am. I would rather you tweet everything onto FB and MySpace than to just close down the profiles and never use them.

**Timing**

You don't have to be on social media all day to be effective. For your followers we know you have a life and a job. We do tend to notice patterns. There are some people who I will see in the morning, others in the evening. Likely if you tweet every day after work, you will make friends with people who are on at the same time.

Just scan the columns on your TweetDeck and go through the three steps—communication, conversation, and reciprocation. Who cares if you encourage someone who hasn't been on Twitter for hours? It will still be meaningful if, when she logs on in the morning, she reads, *Sorry your cat died. Pets are so special, and it is so painful to let them go.* The people who are on Twitter can benefit from a blog you post the next morning too. Many times conversation and reciprocation merge together. Read a friend's blog and then repost. Two steps in one action. Bam!

Then feel free to dive into the conversation.

The great part about approaching social media this way is that you will be actively looking to support and edify others. In turn, they will repay your kindness (more often than not). You will still get what you want—an ever expanding platform—while helping others get what they want—encouragement, interaction, community. You won't even have to ask for their help. People will give it freely because you are one of the good guys.

### Some Etiquette

On Twitter, you will be tempted to send auto-follow messages. Don't. Do you like form letters and SPAM? Why do you think others would like it from us? We aren't Domino's Pizza. There are few things I find more annoying than seeing someone has sent me a direct message and it is basically SPAM. DMs are usually special, and the auto-follow ruins that, thus associating anything that person had to say with a negative emotion.

Just interact with people. That will resonate much better than some message that cost you nothing. Your time costs something and has value.

**There are times for multiple Twitter identities.** I am not contradicting what I said earlier, only qualifying. HootSuite and TweetDeck allow you to manage more than one Twitter ID. In my opinion, only do this if some aspect of what you do requires regular announcements.

For instance, when Bob first started tweeting, he also had just started his Warrior Writer Workshops. Bob was so busy writing, learning all the social media stuff I was throwing at him *while* developing the workshop, that he often forgot to *mention* the upcoming workshops. That was bad for his workshop business in that Twitter, when he used it, was very effective at filling slots.

I recommended an @WarriorWriter Twitter ID that Bob could program to make announcements regularly about upcoming workshops. The difference here is that those following are not going to expect interaction with @WarriorWriter. They KNOW it is purely for updates regarding Bob's workshop schedule. Thus, Bob's followers who wanted to keep up with workshops could choose to follow @WarriorWriter as a reminder to them about workshops, and they could keep an eye for anything that they could take advantage of. They could also be nice and repost them for Bob.

A tweet by @WDWPUB or @Warrior Writer can be programmed to "bot" messages. It won't generally offend anyone unless you post so much you become annoying. Do not try to do this with your regular tweets. No matter how witty and clever you think you are being, we will see through it very quickly and will not like you for being phony-baloney. Genuine interaction goes much farther with us. Don't worry. A little goes a long way.

You always want to strive to keep the numbers of your followers and the number you are following roughly equal. **Make sure you follow everyone who follows you**. If you are using TweetDeck then it won't matter if they have anything you want or even if you find them interesting. If they suddenly become interesting, you can move them to the column with the interesting people. It is just good manners to follow people who follow you. Now if they look like a weirdo or a bot, then don't worry about following them. But for people who look like bonafide twits...tweeters...tweeps, then make sure you reciprocate.

**Your Twitter Karma** is a very useful tool for making sure you are staying on top of following everyone kind enough to follow you. Just type *your twitter karma* in the browser and then follow the cues. I recommend following everyone you can. Your Twitter Karma will tell you all the people who haven't tweeted a lot or even in a long time. You might be tempted to unfollow them. Don't. You don't know why these people are inactive. They might have had a death in the family or are on a deadline for their college thesis. They could be just like you and maybe just got started on Twitter. The fact is, you don't know. Give the benefit of the doubt. They aren't taking up any space, so just leave them be. They could start tweeting one day and end up your biggest fan.

I would avoid following anyone with an overly strange Twitter handle or something that looks like it is just going to be for spamming you. For instance, a name like @john57639405_guy is just weird. What person would pick a name like this? Sorry if I just offended @john57639405_guy (my advice, change your name, hon). But, you guys get the point. Sometimes, it is just a matter of common sense.

Also, avoid using services that verify if people are real people on Twitter. That is like making people solve CAPTCHAs. It is an annoying hoop to jump through, especially on Twitter since it takes all of a half second to unfollow someone if they SPAM you. I can guarantee you that I have yet to follow the cues and validate I am a real person. I just ignored the DM and moved on and asked other people to be part of my network instead. I generally have fifty things going at once and it isn't anything personal, I just choose to befriend people who make it easy to be friends.

### Outsourcing

PR firms and marketing firms are a valuable resource. This book has been about you and your part on social media. Most of us, when we start out, are unsure who we are and what we want. My techniques give you the way to get started and grow so that, by the time you are ready to outsource, your firm of choice has something solid to work with.

Once you reach a certain point of success in your field, you very likely will hire publicists and professional firms to take you to another level. But, by maintaining a personal touch in your interactions, you maximize all the nifty things these firms can do for you on their end.

For instance, I didn't have a web site until very recently. I used my MySpace instead. But, I experienced certain dramatic changes in my professional life that commanded I go to the next stage of my career. Suddenly it became critical for me to outsource. Could I have read a *Building Web Sites for Dummies*? Sure. Actually, tried doing it that way. But then I stopped and did some serious thinking. The entire reason I needed a web site was that I was taking off professionally. The time it would take for me to teach myself how to do my web site or get a friend to build a web site could be better spent working on this book and others. It made far more sense to hire someone.

If and when you choose to outsource is a decision only you can make. But, I do want you to remember that you are basically a small business. Hiring a professional to build your pages is a business expense and a tax deduction. Fun City Social Media built me a custom web site and then created matching Twitter, MySpace and Facebook fan pages that all coordinated. They linked everything together to make it fully optimized then took time to train me and walk me through how to use my new site. They did such a wonderful job, that I insisted they put a clickable icon on my page so that others could find them and hire them.

The funny thing is that this is such a great example of what social media is all about. I don't work for Fun City Social Media. But, I was so impressed with them and their work that I wanted to give others an opportunity to enjoy the same service I loved so much. Renée Grosskreutz, Jeff Posey, and Fred Campos put a lot of time, attention and care into crafting a the perfect page for me and my work. They went above and beyond to not just build a page, but to create something that would make me excited. Their pricing was more than reasonable and I have never had to wait more than a couple of hours for any issue, question or problem. I hope that when you make it to a logical point to outsource you will look them up first.

Getting back to my original point, professional firms are a great extension of what you will create by using my program. Notice I said that they are an *extension*. They are not a *substitution*. There is no substitute for special unique you. Unless you are a huge mega-author, today's reader expects a different level of interaction than years ago. And for the huge mega authors, a little interface will go a long way.

## Conclusion

By now I trust that you are excited about building your platform on social media. I hope that my program has dispersed a lot of fear, confusion and misinformation and offered you a clear path to a promising future. We writers need to always be mindful that we are small business owners at the very least. Marketing is just a necessary component in our enduring success. But I say, if we have to do something, then we might as well make it fun.

I hope everyone who reads this book will connect with me on MySpace, Twitter, and Facebook. Go to my website www.kristenlamb.org. Renée built nifty icons to make it simple to follow me on all of three platforms. Feel free to follow my friends as well. The more the merrier. I also post a weekly blog about writing and social media to help you stay current with all the latest and greatest tools to make your life simpler so you have time to write brilliant books.

Technology is changing faster than we can teach it, and that is why this book addressed namely the content. Platforms and protocols will change, but the human experience is timeless. Being genuine, thoughtful, and focused on serving others is helpful in more areas than just social media.

Just remember: "Excellence can be obtained if you:
...care more than others think is wise;
...risk more than others think is safe;

...dream more than others think is practical;
...expect more than others think is possible."
~Author Unknown

I wish you all the best. Good luck and happy writing.

*Biography*

Kristen takes her years of experience in sales & promotion and merges it with almost a decade as a writer to create a program designed to help authors construct a platform in the new paradigm of publishing. Kristen has guided writers of all levels, from unpublished green peas to NY Times best-selling big fish, how to use social media to create a solid platform and brand. Most importantly, Kristen helps authors of all levels connect to their READERS and then maintain a relationship that grows into a long-term fan base.

Currently Kristen is teaching *Social Media Marketing for Writers, We Are Not Alone–Social Media for the 21st Century Author* at various writer conferences across the country. Stay tuned for a workshop in your area.

Breinigsville, PA USA
29 March 2011

258689BV00004B/21/P

9 781935 712183